DARK MATTERS

RANDY J. NELSON

DARK
MATTERS

*Harmonizing Our Circadian Rhythms
to Optimize Health and Well-Being*

OXFORD
UNIVERSITY PRESS

OXFORD
UNIVERSITY PRESS

Oxford University Press is a department of the University of Oxford.
It furthers the University's objective of excellence in research, scholarship,
and education by publishing worldwide. Oxford is a registered trade mark of
Oxford University Press in the UK and in certain other countries.

Published in the United States of America by Oxford University Press
198 Madison Avenue, New York, NY 10016, United States of America.

CIP data is on file at the Library of Congress

ISBN 9780197639948

DOI: 10.1093/9780197639979.001.0001

Printed by Integrated Books International, United States of America

The manufacturer's authorised representative in the EU for product safety is Oxford University Press España S.A. of El Parque Empresarial
San Fernando de Henares, Avenida de Castilla, 2 – 28830 Madrid (www.oup.es/en or product.safety@oup.com). OUP España S.A. also acts as
importer into Spain of products made by the manufacturer.

To my family. . .

CONTENTS

ACKNOWLEDGMENTS

This book represents significant input from several individuals over the years. First, I thank my family members, Courtney DeVries, Morgan Nelson, and Justin Nelson, who read and critiqued the manuscript. I wish to thank Catherine Carlin, my first and funnest editor for several previous books published at Oxford University Press, who encouraged this project starting nearly a decade ago, and provided input to some early chapters. I hope she sees "funnest" in print. Thanks so much, Catherine. I really appreciate all of your support over the years. Similarly, Sydney Carrol, my textbook editor at Sinauer Press (now Oxford University Press) suggested to me many times over the years to write a trade book on disrupted circadian rhythms. Syd also helped me refine and improve my writing style for this book manuscript with very helpful and kind feedback. Thanks Syd! I met Ada Brunstein when I served as an editor for the *Oxford Encyclopedia of Neuroendocrine and Autonomic Systems.* When she learned about my goal of writing this book she was super supportive. She helped me form the premise and outline of the book, and solicited reviewers who helped form the proposal and direction of this work—thanks so much, Ada! After a well-deserved promotion, she handed this book project to the current book editor at Oxford, Tim Allen. Tim has also been super supportive and has provided much needed guidance in writing this "trade book." We have developed an annual ritual to meet at the Society for

Neuroscience conference, have lunch, and based on my excuses, adjust the completion date. But alas, we'll need a new ritual as we both bask in the relief of completion of the book manuscript. I appreciate his gentle editing—the book is much better and accessible due to his efforts. Thanks a 10^6, Tim! I'm grateful to Molly Thompson for the amazing artistic renditions in the book. Finally, I thank the talented project editor, Chioma Anomnachi, for her help in pushing the manuscript into production.

1

INTRODUCTION TO CIRCADIAN RHYTHMS

I have a good friend, Jennifer, a nurse who works the night shift in a critical care unit. Although she has been a nurse for many years, she volunteers for night-shift work because it pays more than the same job on day shift. She arrives at the hospital around 7:00 p.m. (1900h), and then works until about 7:00 a.m. (0700h). So, when her family and the vast majority of other people are sleeping, she is actively working, eating her meals, and interacting with her colleagues and patients. Although she sometimes goes out for breakfast and beers with her colleagues at a bar that caters to medical staff after the night-shift workers end their shifts, Jennifer generally gets home in time to see her husband and daughter off to work and school. After they leave, she usually makes time to do some chores, and run some errands. Prior to noon, she goes to bed and sleeps—but she revealed to me and my wife that she never sleeps well. She wakes up around 5:30 p.m. to make dinner for her family and prepares for work. In addition to the fitful sleep, Jennifer suffers from a number

Dark Matters. Randy J. Nelson, Oxford University Press. © Oxford University Press (2025).
DOI: 10.1093/9780197639979.003.0001

of ailments, most of which began since she started night shifts. She is considering going off night shift, although she has been saying that for years. She has gained significant weight while working night shifts and developed type 2 diabetes. In addition to diabetes medicine, she is also taking medication for depression and hypertension, and as a medical professional, she knows that she is at risk for several chronic diseases. Because she knows that I study biological rhythms, Jennifer asked me if I thought night-shift work might be affecting her health negatively. In response, I gave her a bunch of articles to read which together suggest that indeed, night-shift work was a significant risk factor for several disorders. Yes, night-shift work is associated with suboptimal metabolism and the onset of type 2 diabetes (covered in Chapter 2). Yes, night-shift work is associated with depressed mood (Chapter 3). Yes, night-shift work is associated with disrupted sleep (Chapter 4). The night shift is also linked to memory problems (Chapter 5), cancer (Chapter 6), and cardiovascular diseases (Chapter 7). I thought others might want to have access to this sort of information, and so I decided to write this book. The primary goal of this book is to bring awareness of the health and well-being consequences of poor circadian hygiene and help readers counter any negative health effects of disrupted circadian rhythms. As we'll see, this could mean improving mood, increasing your cognitive function, or even reducing cardiovascular disease and cancer risks!

Why are night shifts so harmful to health? The short answer is that internal rhythmic bodily functions are disrupted by sleeping during the day and being active at night. These internal rhythms reflect the 24-hour solar days comprising relatively bright light during daytime hours and dark conditions during the night, and these rhythms have been an integral part of life presumably since life began.

Life evolved in the oceans of primordial Earth. Scientists speculate that early life resembled that of current single cell creatures—current examples include amoebas and bacteria—living in the oceans. These creatures rely on seawater to provide them with the nutrients, oxygen, water, and basic electrolytes (e.g., sodium, chloride, and potassium) necessary to sustain their life processes. These simple animals are at the mercy of the composition and temperature of the seawater; small changes in water salinity, for instance, may be fatal. To some extent, the evolution of more complex multicellular creatures—living organisms having more than one cell—has required the compartmentalization of "seawater" within the body, in the form of extracellular fluid. Extracellular fluids include all of the bodily fluids located outside cells and include blood plasma, interstitial fluid (the fluid between the cells), and lymph (the clear fluid that travels in the lymphatic system and comprises white blood cells to fight infections). Essentially, these fluids represent a remnant of the seas in which our single cell ancestors evolved—albeit a markedly diluted remnant—necessary to bathe our cells. In humans and fellow mammals, homeostatic, or stabilizing, mechanisms have evolved to maintain the composition and temperature of this extracellular fluid at a relatively constant level despite the changing conditions outside of our bodies.

The prototypical human-made homeostatic device is a thermostat that controls the heating and cooling system in your home. As temperatures decrease below the thermostat setting (e.g., 68°F/20°C) the furnace begins heating, which in turn raises the temperature of the area controlled by the thermostat. When the room temperature rises to the "set" temperature, the thermostat will turn off the furnace. If the ambient temperature rises far above the set temperature, then the cooling system will be activated, turning off when the area becomes too cool. The thermostat acts to keep the room temperature within a relatively narrow range around the set point. The deviation in room temperature from the set point is typically just a few degrees before the furnace or air conditioner is activated or turned off; for example, the thermostat set at 68°F/20°C may keep the room between 65.3°F/18.5°C and 71°F/21.5°C. Otherwise, the system would be continuously turning on and off in response to minor changes in temperature. Conceptually, this regulation is similar to homeostatic processes in the body that maintain values such as pH, blood sugar, and body temperature within very narrow ranges.

For example, the processes regulating intake and excretion of water and sodium, the two main components of our extracellular fluids (the bodily fluids that bathe our cells), are closely linked so that they are maintained at ideal levels within relatively narrow ranges. The ability to maintain equilibrium between the internal water and sodium environment, despite quite variable external conditions, has allowed multicellular animals to inhabit almost every niche on the planet. The maintenance of a relatively constant internal environment liberated animals from the sea, or as Claude Bernard stated, allowed *la vie en liberté*, the free life.[1]

Although the notion that our bodies work to maintain homeostasis permeated biology and medicine for decades, it turns out that virtually all bodily functions vary throughout the day (Figure 1.1). This temporal variation reflects bodily processes of

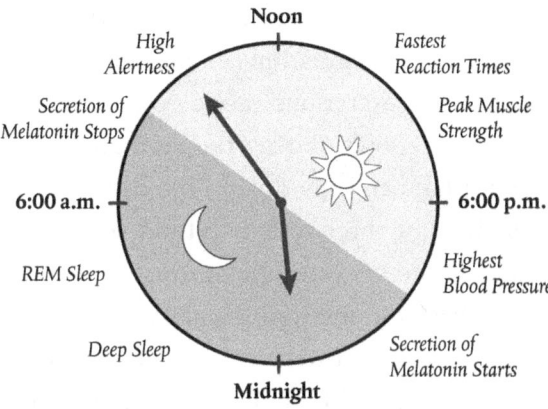

Figure 1.1 Circadian rhythms in physiology and behavior. Examples of biological processes that occur at specific times of day in humans. These rhythms are regulated by the primary biological clock in the suprachiasmatic nuclei (SCN) in the hypothalamus. Peak and base (nadir) circadian timing of some of these and other physiological and bodily functions are depicted here (BP = blood pressure).

primitive life forms that may have occurred in response to the time of day. These are regulated by homeostatic processes, but the setpoints vary throughout the day. You can think of these processes as being regulated by a more sophisticated thermostat that can be programmed to limit heating or cooling to times when people are likely to be home. This temporal (time-of-day) variation increases the efficiency of the home heating system. Similarly, biological timing systems increase biological efficiency and thus increase fitness by changing the setpoints of all physiological processes throughout the day and night.

These processes require energy, and there simply is not enough energy available to do everything all the time. Thus, many biological processes became compartmentalized to occur only during certain times of day or were optimized at certain times of the day. For example, osmoregulation (maintaining water balance), pH, and even bodily excretions vary across the day. Some processes could happen only during the sunlight, whereas other processes could happen only during the dark of night. As life evolved over the past three to four billion years, creatures not only internalized the seawater for optimal biological function but also internalized the environmental pattern of light days and dark nights created by the Earth's daily rotation. Our bodies have become so exquisitely calibrated to this 24-hour pattern that, in common with our unicellular ancestors, we conduct certain biological processes only at night and others only during the day. These processes reflect internal biological manifestations of the solar day that are approximately, not precisely, 24 hours. For example, blood pressure, body temperature, and sleep are common rhythmic functions—even bowel movements are typically suppressed after about 10:00 p.m. for most folks. Such rhythms are referred to as circadian rhythms and these rhythms are driven by circadian clocks. The term circadian is from the Latin, *circa*, which means "approximately" or "about" and *diem*, which means "day." Typically, some physiological processes are more pronounced or optimized at one time of day compared to another (Figure 1.1).

Although we have learned a lot about circadian clocks and rhythms over the past few decades, turning this foundational knowledge into clinical applications or solutions to health-related problems has been slow. This could be caused by the resistance of 20th-century scientists and doctors to acknowledge that substantial fluctuations in physiological processes were not merely pathological (caused by a physical or mental disease), but were related to programmed changes in physiology and behavior that we now know underlie homeostatic processes. In hospital patients, for example, the programmed elevation of body temperature in the early evening was often considered an evening "fever," and treated with aspirin. Additional barriers to incorporating circadian rhythmicity into the study and treatment of diseases are based on practical considerations, such as patient compliance with medication dosing times, and additional cost incurred by including multiple dosing times in clinical trials.[2] Indeed, the effectiveness of many medications displays daily rhythms, but many physicians remain unaware of this variation. Despite the recent awarding of a Nobel Prize to prominent circadian biologists, Jeffrey C. Hall, Michael Rosbash, and Michael W. Young, in 2017 and a growing literature describing beneficial effects of differential timing of chemotherapy, anesthesia, and drug effectiveness, translation to clinical practice remains virtually nonexistent. As a result, this indifference to circadian rhythms by the medical professions has diluted the importance of these temporal cycles to most of us. As we will see over the course of this book, disruption of these internal daily rhythms through seemingly innocuous activities, such as watching TV or using our smart phones at night, can negatively affect this temporal dynamic and alter how our internal clock system regulates our bodily functions and behavior.

Virtually all organisms on the planet have self-sustaining, internal biological clocks with periods of ~24 hours. These clocks likely evolved to anticipate light and dark so that plants could efficiently orchestrate photosynthesis or counteract damaging redox reactions on a daily basis. Redox reactions, which are important in producing and transferring energy from food to power our cells, can be both positive and negative, and must be balanced for optimal bodily function. Negative redox reactions include oxidative stress (which is associated with various diseases such as cancer and diabetes) and aging. As organisms became more complex, these internal clocks regulated not just metabolism, but other functions as well. In humans, virtually every aspect of our physiology and behavior, including sleep, hormone secretions, body temperature regulation, cell division, immune function, metabolism, and food intake, is mediated by our internal biological clocks.

Again, these internal rhythms synched to the solar days are called *circadian rhythms*. The "biological clocks" that generate circadian rhythms are called circadian clocks. In mammals, this primary biological clock consists of a cluster of 20,000 nerve cells located at the base of the brain in the hypothalamus (Figure 1.2). This paired structure is called the suprachiasmatic nuclei, but I will simply refer to it as the SCN.

This clock obtains light information from the environment regarding the solar day and conveys it to the rest of the body by either humoral (blood-borne) or neural signals (Figure 1.3). Among mammals, a nonvisual neuronal pathway, the retinohypothalamic tract, conducts light information from non-image-forming retinal ganglion cells directly to the SCN.

These retinal ganglion cells comprise the photopigment, melanopsin; melanopsin is most sensitive to short wavelengths of visible light (i.e., blue) and least sensitive to long wavelengths of light (i.e., red) (Figure 1.4). Many light-emitting devices are enriched with blue light. Thus, your TVs, laptops, phones, and eReaders when used at night can disrupt your circadian rhythms.

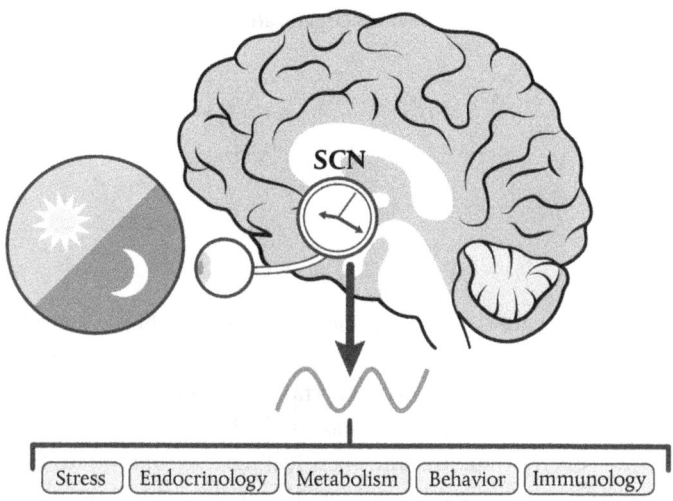

Figure 1.2 The SCN are the central biological clock that regulates circadian rhythms. Light information is transmitted from our eyes to the SCN at the base of the brain. These signals are separate from those transmitted from the eyes to the visual cortex in the brain. Importantly, the receptors in our eyes that transmit light information to the primary clock in the brain are maximally sensitive to short wavelength (blue) light. Exposure to such light sets our internal circadian rhythms that run with a period of *approximately* 24 hours to *precisely* 24 hours (the solar day). This rhythmic information is transduced to the rest of the body via chemical and neural signals so that our bodily functions, such as stress responses, endocrine function, metabolism, timing of sleep and other behaviors, and immune function, are synchronized.

If individuals are left in constant conditions, such as a dark cave, then the internal circadian rhythms emerge—and they are not exactly 24 hours—hence the wiggle room in the term "circadian." For example, if you were placed in a dimly lit cave for several weeks, your sleep-wake cycle might stretch out to about 24 hours and 15 minutes, just slightly longer than the 24 hours it would be above ground, where the circadian clock is being reset

9

Circadian Organization

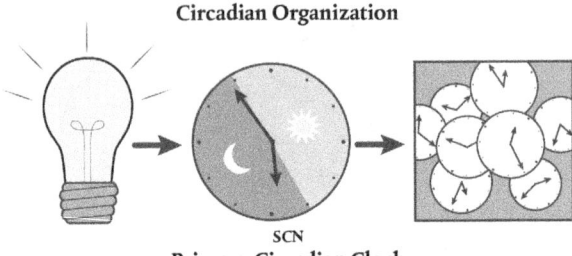

SCN
Primary Circadian Clock

Figure 1.3 There are three components to the circadian system in people. (1) The input system comprises special photopigments in non-image-forming cells in the eye that transmit light information to the (2) primary circadian clock at the base of the brain, the suprachiasmatic nuclei (SCN). (3) Both neural and chemical signals from the SCN provide temporal information to all the tissues throughout the body to organize and coordinate physiology and behavior so that it is aligned with the solar day.

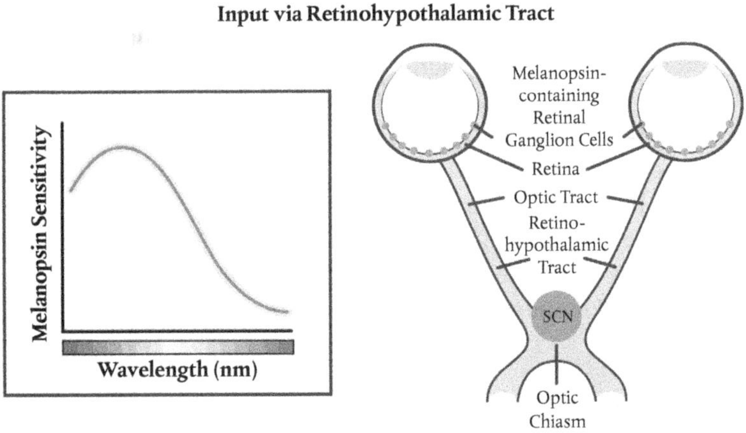

Input via Retinohypothalamic Tract

Melanopsin Sensitivity

Wavelength (nm)

Melanopsin-containing Retinal Ganglion Cells

Retina

Optic Tract

Retino-hypothalamic Tract

SCN

Optic Chiasm

Figure 1.4 Light input is via the retinohypothalamic tract that travels along the optic nerve. Light is transduced to the SCN via non-image-forming ganglion cells in the retina via the retinohypothalamic tract. The photopigment in these cells is melanopsin. Melanopsin is maximally sensitive to short wavelength (blue) light.

daily. Other rhythms such as body temperature and digestion, as well as cortisol (stress hormone), testosterone, and estrogen (sex steroid hormones) secretions may also display rhythms that are not precisely 24 hours, and also not aligned with one another. Likely, you've experienced these so-called free-running circadian rhythms if you have traveled on a jet across several time zones. Travel across six or more time zones often leads to a feeling of malaise termed jet lag; jet lag will be discussed in more detail later in this chapter. If you are exposed to light during your biological night, whether due to jet travel, or night-shift work, or watching movies late at night on your electronic device, then your clocks will be inappropriately adjusted, and your body will not function optimally.

Some biological clocks are set by non-light factors. For example, there are biological clocks that are set (entrained or synchronized) to exactly 24 hours by food intake. These so-called food entrainable oscillators (timers) are responsible for setting the 24-hour rhythms of food anticipatory activity. Unlike the central clock located in the SCN, the underlying mechanisms of circadian food entrainment and precisely where this process occurs remain unspecified.[3] Another series of circadian clocks that are not mediated by the SCN are the redox cycles that are important for rhythms in metabolism and potentially immune function.[4] We will return to food entrainable oscillators later.

At a molecular level, circadian rhythms are generated by self-regulatory cycles of circadian clock genes being activated to drive subsequent protein production. As these so-called clock proteins build up in the cell, they feed back to the genetic machinery to inhibit further gene transcription (Figure 1.5). This cycle of gene

Endogenous Oscillator: Suprachiasmatic Nuclei (SCN)

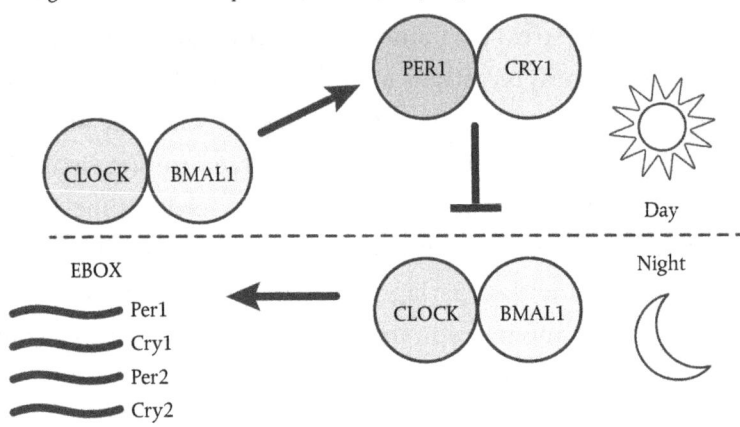

Figure 1.5 Circadian rhythms are generated by the autoregulatory transcriptional feedback loops of the core circadian genes. In the nuclei of SCN neurons, BMAL1 and Clock, two so-called circadian clock proteins, form a new molecule called a heterodimer that binds to E-box sequences in the promotors of the *Cry* and *Per* genes to activate gene transcription at the beginning of the circadian day. The gene products of *Per* and *Cry* accumulate, dimerize (pair together), and then form a complex with another molecule that moves back into the nucleus to interact with CLOCK and BMAL1, repressing their own gene transcription. This feedback cycle takes about 24 hours. Exposure to light early during the early part of the circadian day sets these cycles to exactly 24 hours.

transcription takes approximately 24 hours. These cycles are set to precisely 24 hours by exposure to bright light early during the day.

Indeed, for most circadian rhythms, exposure to bright light during the day, especially early in the day, resets the circadian clocks to precisely 24 hours each day. Exposure to dark nights is also important for good circadian hygiene. However, exposure to artificial light at night can derail this system and wreak havoc with the temporal coordination of physiology and behavior.

Figure 1.6 A global view of earth at night. This composite image was compiled from over 400 satellite images depicting light pollution across the globe. Reproduced from nasa.gov (2017). Earth at Night. NASA/NOAA. https://www.nasa.gov/topics/earth/earthday/gall_earth_night.html.

Exposure to light at night is pervasive in the modern world (Figure 1.6). A recent large-scale study investigating worldwide patterns of light pollution demonstrated that anthropogenic (human-produced) sky glow dominated over celestial light. Indeed, the Milky Way is hidden from view by light pollution for 80% of North Americans and 60% of Europeans. If you wish to see pristine skies, then visit Chad, the Central African Republic, or Madagascar. These areas of the planet have the least amount of nightly light pollution. Thus, light pollution has eliminated dark nights in many areas, which has widespread ecological consequences, as well as likely health consequences for people and our companion animals.

Disruption of circadian rhythms is a common occurrence for most people on the planet. The health consequences of misaligned biological rhythms are only now being realized. Some forms of circadian rhythm disruption are subtle, such as exposure to light at night. Other forms of disrupted circadian rhythms can be dramatic, such as night-shift work or jet lag. Again, exposure to light early in the day adjusts your internal clock to precisely 24 hours. Ideally, exposure to outside light during the first hour after dawn or first hour after awakening sets your clock to 24 hours. Let me repeat the core message of this book: If you are exposed to light during your biological night, whether due to jet travel, or night-shift work, or watching movies late at night on an electronic device, then your clocks will be inappropriately adjusted, and your body will not function optimally!

In the next sections, I will review the various forms of disrupted circadian rhythms. In each of the following chapters, this organization of the various forms of circadian rhythm disruption—namely, jet lag, social jet lag, exposure to dim light during the day and exposure to light during the night, and night-shift work—will be repeated to understand their effects on various aspects of health.

Jet Lag

As mentioned previously, the temporal niche of humans during the past million or so years has remained quite stable, with the Earth's rotation time slowing down by only about 20 seconds. In contrast, the technological changes of the past century have produced dramatic and unprecedented effects on our temporal environment. The development of jet travel has led to abrupt phase shifts between our internal biological rhythms and the solar day not previously encountered by travelers. Most jet travel occurs on east–west, rather than north–south, routes because many major commercial centers in Asia, Europe, and North America are at similar latitudes. Most people have likely experienced jet lag, which amounts to the physiological and behavioral responses to travel across time zones. Jet lag involves phase shifts in all the *zeitgebers* (this is from the German term, "time giver," which refers to the external stimuli that synchronize internal circadian rhythms to the environmental time) at once. Symptoms of jet lag include (1) sleep disruption; (2) disruption of metabolic processes including digestion; (3) impaired behavioral function, including attention, perception, and motivation; and (4) a general feeling of malaise. Generally, the severity of jet lag symptoms correlates with the number of time zones crossed.

Virtually everyone can adapt relatively easily to 1-hour phase shifts associated with the changes between Daylight Savings Time and Standard Time, but there are substantial individual differences in how temporal phase shifts affect performance, as well as in the speed of resynchronization following time shifts. Generally, elderly people have more difficulty with the effects of jet lag than younger people. Chronic jet lag probably only affects a few individuals, such as flight crews or frequent international flyers, but there have been studies of chronic jet lag on health, and these will be described in subsequent chapters.

Social Jet Lag

Although chronic travel-related jet lag is experienced by relatively few individuals, social jet lag is experienced by virtually all of us. Remarkably, we all experience social jet lag from preschool until retirement, and have likely never thought about it! Social jet lag is often a weekly phenomenon experienced because of early awakening on work and school days and then staying up late and consequent late awakening on days off. Social jet lag is pervasive and causes a perpetual back-and-forth "jet lag" of shifting rhythms that may compromise health. For example, you may stay up three or more hours later on Friday and Saturday nights and sleep in later on Saturday and Sunday mornings. This is equivalent to flying from the East Coast of the United States to the West Coast, and then back every week! The extent to which social jet lag influences health by disrupting circadian rhythms or disrupting sleep, an important set of "hands" on the circadian clock, is just beginning to be understood.[5] However, the common prevalence of social jet lag as a modern disrupter of circadian rhythms suggests that more research is necessary.

Exposure to Dim Light During the Day

Many of us who work indoors or go to school all day or live in assisted living facilities rarely venture outside. For the elderly, this can lead to a lack of a strong light signal to set their circadian rhythms to precisely 24 hours each day, causing their physiological functions to drift out of synchrony. Although there is significant variation among individuals regarding how much light is necessary during the day to maintain optimal circadian function, for the vast majority of folks indoor lighting is insufficient. Thus, for many people living in assisted living facilities, staying in hospitals, working in offices, or attending primary or secondary schools, the lack of exposure to outdoor lighting is comparable to living in a cave: our circadian rhythms can drift out of synchrony, which leads to suboptimal health. To maintain optimal health, exposure to relatively bright light during the day and dark during the night is ideal for circadian regulation. Most of us require at least 1,000 lux (about 5 times the light levels of most indoor rooms and equivalent to an overcast day) or more of light exposure during the day to set our clocks to 24 hours. You can see some common light levels in Figure 1.7.

Approximately 90% of our time is now spent indoors[6] under electric light that is typically 100 times dimmer during the day than natural daylight, and 1,000 times brighter after dusk than even the brightest moonlight. This inverted light-dark schedule causes significant health problems, as noted throughout this book. For example, the World Health Organization (WHO) International Agency for Research on Cancer (IARC) classified night-shift work with circadian disruption as a probable human carcinogen in 2007.

It is especially challenging during autumn and winter, when day lengths are relatively short in the Northern Hemisphere, to

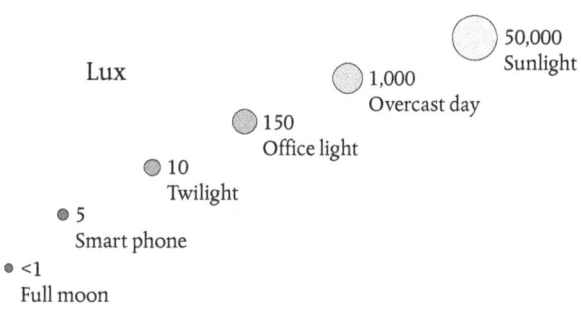

Figure 1.7 Relative illumination levels (lux).

get outdoors for bright light exposure. Dysregulation of circadian clocks can lead to a significant mood disorder called seasonal affective disorder (SAD). SAD is characterized by depressed mood, lethargy, loss of libido, increased sleep, excessive weight gain, carbohydrate cravings, anxiety, and inability to focus attention or concentrate that occur during the late autumn or winter. In the Northern Hemisphere, SAD symptoms usually begin between October and December and go into remission during March. These symptoms do not merely reflect the holiday blues—individuals suffering from SAD in the Southern Hemisphere display symptoms 6 months out of phase with Northern Hemisphere residents. With the onset of summer, SAD patients regain their energy and become active. Three features atypical of depression—increased food intake (hyperphagia), carbohydrate cravings, and hypersomnia—set SAD apart from nonseasonal depression. SAD prevalence rates in the population range from 1% to 10%, with higher prevalence rates reported at higher latitudes. Circadian rhythms that are not appropriately synchronized with the light-dark cycle may contribute to SAD. Indeed, the standard treatment of SAD is with bright light therapy. When patients are exposed to bright light (~2000 lux), typically for a few hours in the morning, reduction in SAD symptoms is often apparent within a few days. Bright light treatment in the morning may ameliorate depression by realignment of inappropriately entrained circadian rhythms. Light treatment at different times during the day results in differing rates of mood improvement; light treatment during the evening has no mood benefits.[7] Light (phototherapy) therapy appears to shift circadian rhythms by altering the timing of the nightly secretion of melatonin, a hormone released from the pineal gland in the brain.

We will explore SAD in more detail in Chapter 3, and we will explore the role of disrupted circadian rhythms and melatonin in more detail throughout the book.

Late Night Food Intake

In 1955, an interesting new medical syndrome was reported in the scientific literature, called night eating syndrome (NES). This disorder was characterized by anorexia during the morning, increased night-time food intake, and insomnia. People with this condition often wake from sleep to eat. Unlike night-shift workers, the timing of the wake-sleep activities is largely typical. Thus, individuals with NES allow us to separate out the effects of night-time food intake from other aspects of night-shift workers' schedules. It has become clear over the past two decades that meal patterns and timing of food intake affect metabolism. NES leads to obesity and increased risk for developing type 2 diabetes, likely reflecting the observation that our bodies do not expect food at night and our tissues are insulin-resistant during this time. Indeed, nearly half of obese individuals (i.e., those selected for bariatric surgery) fit the clinical criteria for NES; most of these individuals present with type 2 diabetes. Also, individuals with NES present with decreased melatonin concentrations. Timed melatonin treatment of individuals with NES helps to improve the timing of food intake and reduce the individual's body weight. Conceptually, NES is the opposite of intermittent fasting diets, which promote weight loss by restricting food intake between sunset and dawn. More details will be provided in Chapter 2.

Although once considered relatively rare, the results of a recent study of a cross-section of a Greek population (533 individuals) suggests that prevalence of this disorder may be increasing.[8] A little over 8% of this population experienced NES symptoms.

Exposure to Light at Night

Nonhuman animal studies have reported that exposure to very low levels of light at night (5 lux—about the amount of light from a child's night light 1-2 meters from your eyes), causes a host of problems for these rodents including increased risk of some types of cancer, impaired immune function and metabolism, and changes in cognition and mood. It appears that changes in the molecular clock mechanisms are altered by this nighttime light exposure, which causes widespread dysregulation of temporal coordination and optimal health. The full extent to which light at night affects our sleep remains generally unspecified, but recent studies from Northwestern University indicate significant effects of inside lighting on our sleep.[9] In any case, it is well established that optimal brain health and function depends on optimal sleep quality and quantity.

Of course, it is important to practice good sleep hygiene. However, exposure to light at night, as well as lack of exposure to bright light during the day can influence circadian organization, and sleep is regulated by this circadian system. Most people in modern societies differ from our ancestors in two important ways: (1) they are not exposed to sufficient bright light during the day and (2) they do not experience true dark at night. Serious health issues seem to be linked to this departure from our ancestral light and dark environmental exposures. Indeed, exposure to low levels of daytime lighting is also implicated in the disturbance of circadian rhythms and the sleep-wake cycle. A recent study on lighting environments in the workplace demonstrated that workers in windowless environments experience poorer sleep quality, shorter sleep duration, and more frequent sleep disturbances (as assessed by self-report and actigraphy [measurement of movement] recordings) than people experiencing typical outdoor daytime light levels.

Importantly, the sensitivity of the circadian system is tuned toward shorter wavelengths (bluish light), specifically in the range of 460–480 nm. Sunlight reaching Earth is composed mostly of short wavelengths during midday, but those wavelengths become scattered at dusk, when the sun approaches the horizon, causing red wavelengths to predominate. Artificial lights produce different spectrums, depending on the type, but virtually all of our electronic devices (TVs, computer screens, smart phones, tablets, eReaders) emit mostly short wavelength light. Incandescent bulbs emit more peaks in the long wavelength (red) spectrum than in the green and blue spectra, whereas gas discharge lamps emit peaks in the shorter wavelengths. Choosing a lighting system with an appropriate spectral distribution, or filtering light to achieve such a distribution, can minimize disruption to the circadian system. For example, broad-spectrum light is appropriate for daytime indoor lighting, whereas using reddish lighting at night may be beneficial for humans and other mammals.

In one study, it was reported that people who wore lenses to filter out blue wavelengths during simulated shift work normalized their melatonin (a hormone secreted exclusively at night) rhythms compared to people without the lenses. Even in the home, spectral exposure can now be conveniently managed using products designed for this purpose. For example, "smart" LED lightbulbs can be programmed to adjust intensity and wavelength based on the time of day. In addition, there are free apps for smartphones and tablets that adjust the light emitted from the screens to red in the evening. In another study, adolescents who wore blue-light-blocking lenses while using electronic media before bedtime were protected from the melatonin suppression and increased nighttime alertness associated with these activities. People exposed to amber (long wavelength) light for an hour prior to a cognitive test did better than individuals exposed to blue light for the same period of time. Of course, in the ideal scenario, daytime light exposure includes bright, natural light, and nighttime light exposure is minimized altogether.

Night-Shift Work

It is virtually impossible to do long-term experimental manipulations of the light-dark cycles on people. However, we do have some preliminary ways to examine the effects of circadian rhythm disruption on health. For example, night-shift workers are exposed to light at night and often are not well-synchronized to their night shifts—that is, are often not fully synchronized to their reverse light-dark cycles. Thus, they are exposed to light during their biological night, have disrupted sleep, and are eating at suboptimal times of day. In many cases, medical workers, police, and other first responders, for example, experience dim light during their night work shifts, further compromising their circadian rhythms. Night-shift workers suffer from disturbed circadian and sleep rhythms for most of their occupational lives due to exposure to artificial light cycles and the interference of work hours with traditional sleep timing. Ailments that disproportionately plague the night-shift working population are frequently used as a barometer of physiological consequences of circadian disruption. In a way, night-shift workers are inadvertently serving as the "canaries in the coal mine" to warn us about the dangers of disrupted circadian rhythms. Epidemiological studies indicate that night-shift workers display higher incidence of several types of cancers, including breast, colon, and testicular cancers, as well as mood (depression) disorders, cardiovascular diseases, and metabolic disorders compared to their colleagues working the same job on the day shift. We will return to the effects of night-shift work on these and other disorders throughout the book.

We spent centuries battling the night's dangers: predators, crime, fire, and even the idea of ghosts. It makes sense, then, that we eagerly flooded the night with bright electric lights as soon as the technology became available. Unlike earlier ways of creating light—torches, fires, oil lamps, candles—electric lights were a dramatic change: safer, cheaper, more reliable, and easier to use. We adopted electric lights long before we understood circadian biology, so we did not know that we should consider the problems that bright artificial light at night could cause for our mental and physical health. In retrospect, it is perhaps not unexpected that there are biological consequences to radically changing from the predictable light-dark cycles under which life evolved for billions of years to man-made cycles, manipulated by artificial light.

Data on people remain mostly correlational, and the direct role of light at night remains to be established conclusively in humans. However, there is no doubt that healthy sleep is necessary for optimal health, and sleep is regulated by our circadian rhythms. It seems prudent to curtail exposure to blue light during the night and maximize exposure to blue light during the morning in concert with what our ancestors experienced.[10]

"Paleo Lighting"

You have likely heard about the Paleo diet. Essentially, it is based on eating only the types of foods assumed to have been eaten by early humans during the Paleolithic or Stone Age period (which occurred about 2.5 million years). This diet comprises mainly meat, fish, vegetables, and fruit. It has been asserted that this high protein/low carb diet is in harmony with the metabolic adaptations that evolved in humans. Although there are several caveats to this type of diet (for example, hunting and eating wild meat is vastly different from barbecuing modern domesticated meat purchased at Costco), overall, the benefits of the Paleo diet seem to outweigh the risks for most people. One consistent theme of this book is to encourage readers to adopt a "Paleo lighting" regime. It is a simple strategy for harmonized, good health: Go outside of your house or office early in the day to seek outdoor light and avoid exposure to artificial light at night. Although usually sufficiently bright for reading, indoor light levels are typically between 200 and 500 lux, whereas outdoors light levels at mid-day range between 25,000 and 100,000 lux, depending on season, latitude, and cloud cover. Full moonlight may approach 1 lux, whereas a tablet held up to your face typically is between 30–50 lux (Figure 1.7)—at least 30 times brighter!

Again, the primary goal of this book is to bring awareness of the health and well-being consequences of poor circadian hygiene and help you counter any of the negative health effects of disrupted circadian rhythms. The chapters in this book will bring together the most reliable research on the different ways exposure to light at night affects how our bodies function and, ultimately, our health. In the following seven chapters, I will explain how light at night negatively affects several physical, behavioral, and mental conditions. As mentioned (but critically important so it bears repeating), artificial light tends to comprise short (blue) wavelengths. This means that exposure to artificial light at night can derail our circadian clock and wreak havoc on the coordinated timing of our physiology and behavior, leading to impaired metabolism, altered cell division cycles, as well as compromised immune, brain, and heart function. I will cover each of these topics in the next six chapters. In Chapter 2, the effects of disrupted circadian rhythms on body weight is described. Chapter 3 focuses on the effects of light on mood. The effects of light at night on sleep, important to my friend on night shifts, is discussed in Chapter 4. The role of light and misaligned circadian rhythms on cognition and memory is reviewed in Chapter 5. Chapter 6 focuses on the role of light at night on cancer, and Chapter 7 focuses on cerebrovascular issues after exposure to light at night.

As I will reiterate throughout, all these significant health effects arise from disrupting one subtle mechanism, our circadian clocks. Although I will provide general information to avoid disrupted circadian rhythms throughout the book, in the final chapter (Chapter 8), I summarize some general strategies to offset the negative consequences of disrupted circadian rhythms. Importantly, before adopting any of the suggestions for improving circadian hygiene always discuss specific strategies with your healthcare professional.

References

1. Bernard C. 1856. *Leçons de physiologie expérimentale appliquée à la médicine faites au Collège de France*. Vol. 2. Baillière, Paris.
2. Walton JC, Walker WH, Bumgarner JR, Meléndez-Fernández OH, Liu JA, Hughes HL, Kaper AL, Nelson RJ. 2021. Circadian variation in efficacy of medications. *Clinical Pharmacology and Therapeutics*, 109: 1457–1488.
3. Pendergast JS, Yamazaki S. 2018. The mysterious food-entrainable oscillator: insights from mutant and engineered mouse models. *Journal of Biological Rhythms*, 33: 458–474.
4. Reddy AB. 2016. Redox and metabolic oscillations in the clockwork. In: *A Time for Metabolism and Hormones*, P Sassone-Corsi, Y Christen (Eds.), pp. 51–61. Springer Nature, Cham (CH).
5. Caliandro R, Steng AA, van Kerkhof LWM, Van der Horst GTJ, Chaves I. 2021. Social jetlag and related risks for human health: a timely review. *Nutrients*, 13: 4543.
6. Klepeis NE, Nelson WC, Ott WR, Robinson JP, Tsang AM, Switzer P, Behar JV, Hern SC, Engelmann WH. 2001. The National Human Activity Pattern Survey (NHAPS): a resource for assessing exposure to environmental pollutants. *Journal of Exposure Analysis and Environmental Epidemiology*, 11: 231–252.

7. Lewy AJ, Bauer VK, Cutler NL, Sack RL, Ahmed S, Thomas KH, Blood ML, Jackson JLM. 1998. Morning vs evening light treatment of patients with winter depression. *Archives of General Psychiatry*, 55: 890–896.

8. Blouchou A, Chamou V, Eleftheriades C, Poulimeneas D, Rigopoulou E, Bogdanos DP, Goulis DG, Grammatikopoulou MG. 2024. Beat the clock: assessment of night eating syndrome and circadian rhythm in a sample of Greek adults. *Nutrients*, 16: nu16020187.

9. Mason IC, Grimaldi D, Reid KJ, Warlick CD, Malkani RG, Abbott SM, Zee PC. 2022. Light exposure during sleep impairs cardiometabolic function. *Proceedings of the National Academy of Sciences*, 119: e2113290119.

10. Smolensky MH, Lamberg L. 2000. *The Body Clock Guide to Better Health*. H. Holt, New York.

2

LIGHT AND BODY WEIGHT

In 1962, Michel Siffre, a French geological scientist and chrono-biologist (a person who studies biological rhythms), spent two months living in an underground cave, devoid of all natural light and temperature variation. His goal was to determine whether humans possess a biological clock. Indeed, during his time underground, Siffre's sleep-wake habits and mealtimes showed an approximate, but not precisely, 24-hour pattern. For Siffre, his daily sleep-wake, mealtime, and temperature rhythms tracked closer to 24.5 hours. His self-experiment, along with many others that followed, established that humans maintain internal ~24-hour biological rhythms even in the absence of any apparent environmental cues.[1,2] This internal production and maintenance of rhythms is a defining feature of the circadian system. Importantly, circadian organization of our bodily functions is required for optimal performance and health.

Over the course of the 20th century and into the 21st century, the prevalence of obesity and metabolic disorders has rapidly increased worldwide. In 2000, the number of adults with excess

Dark Matters. Randy J. Nelson, Oxford University Press. © Oxford University Press (2025).
DOI: 10.1093/9780197639979.003.0002

body fat surpassed the number of adults who were underweight for the first time in human evolutionary history. Indeed, the growth in obesity has been exponential in recent decades, particularly for the highest weight categories (www.cdc.gov/nchs/products/databriefs/db360.htm). For example, from 2000 to 2005, the number of individuals in the United State with a body mass index (BMI) over 50 (obesity is defined as BMI >30) increased by 75%. Obesity is a pathogenic metabolic condition defined by the accumulation of excess fat (adipose) tissue and is associated with serious health issues including diabetes, cardio-vascular disease, hypertension, asthma, cancer, and reproductive dysfunction. Obesity reduces quality of life, results in significant health-related complications, and more than doubles health-care costs. In addition to traditional risk factors contributing to obesity (e.g., high-calorie diet, sedentary lifestyle, etc.), other environmental factors are likely involved in the development and maintenance of this condition.

What might be these environmental factors? Many historic human innovations have occurred during the past century, such as advances in travel and communication, greater urbanization, and the eradication of several diseases.[3] One environmental change that had a particularly dramatic effect on human lifestyle was the widespread adoption of electrical lighting. Electric lights provide many societal advances. Brightening the night has shed the negative stigma of night as a time solely for crime, sickness, sleep, and death.[4] The use of electric lighting at night played a major role in the Industrial Revolution by allowing for the creation of night-shift work. Furthermore, electrical lighting has given individuals the ability to self-select their sleep-wake schedule. Because the invention of electrical lighting occurred before our understanding of circadian biology, little concern was given to how exposure to unnatural light schedules might affect human mental and physical health. It is reasonable to suggest that exposure to electric light and other devices affects physiology and behavior because of the importance of the circadian system in regulating our homeostatic functions. Think of homeostatic processes as a thermostat—a device that maintains the temperature of our house within a few degrees of the setpoint. The contribution of circadian rhythms is similar to a fancy programmable thermostat that might reduce the house temperature during the day (when no one is home), then increase it prior to the occupants' returning from school and work. Nonetheless, whether the setpoint is relatively low or high, homeostatic mechanisms are integrated with circadian rhythms to program our physiological systems to function optimally.

As noted in Chapter 1, I am suggesting that readers adopt a Paleo lighting regime for improved metabolism, as well as other aspects of health. Imagine the life of an individual human during the Paleolithic period. Without artificial lighting, humans hunkered down, sleeping all night.[4] This was a time to rest and digest. At the start of the day, the first priority of our ancestors was to break their night-long fast (i.e., find and eat breakfast!). In response to a signal from the circadian clock, a hormone, cortisol, increases in our bloodstream in anticipation of waking. Cortisol is often referred to as a "stress hormone," but cortisol is also a key metabolic hormone. Our Paleo ancestors would finish their food intake during the day, and any extra calories not needed to fuel the body during the nightly rest period were floating in the blood as sugar (glucose). These excess calories were stored either as fat in adipose (fat) cells or as glycogen (a version of glucose stored in the liver or muscles) to be used the next morning when activities resumed. Glycogen is made up of many connected glucose molecules and is more readily available as energy than fat. An important function of the morning elevation in cortisol concentrations is to generate and elevate blood glucose by breaking down stored glycogen. This early morning elevation of blood glucose provides the energy necessary to hunt and gather food that is needed to power the body until additional calories are ingested. There is a strong circadian rhythm in blood cortisol concentrations with peak values observed in the morning prior to waking and the start of activity. Metabolism is ramped up at this time and in the unlikely scenario that excess calories are consumed that are not needed to obtain additional food, then secreted insulin pushes these calories into storage for ensuing days. Throughout the day, people were exposed to light levels many-fold higher than modern humans see indoors.

As the day winds down, any excess calories are stored as fat to start the cycle anew. Our bodies are not programmed to expect calories after dark, and, as we will see later in this chapter, any calories consumed at night are almost always converted to fat. Our cells, not expecting an infusion of glucose at night, become temporarily insulin resistant[5]—a physiological state that can become permanent and lead to type 2 diabetes. We may be able to reduce the risk of this fate simply by mimicking the Paleo lighting regime.[6] That is, exposing ourselves to bright light during the day and avoiding light at night will keep insulin resistance at bay. Eat during the day and avoid eating at night!

Disrupted circadian rhythms can upset our metabolic balance and make it easier for us to gain weight. The precise temporal rhythms of physiology and behavior can become dysregulated by modern external light cues such as dim light during the day or artificial light at night, and our health can suffer. Exposure to dim light at night can even affect body mass in children.[7]

Although we have all learned that to lose body weight we need to go on a restrictive diet to eat fewer calories than we expend, dieting fails to result in a long-term significant weight loss for most people. Why is this? Because (1) it is critical to have intact functional circadian rhythms for optimal metabolism and (2) the time when the calories are consumed matters in weight management as much as the number of calories consumed. In the next sections we will explore common disruptors to circadian rhythms and their effects on metabolism and body weight. Spoiler alert: In all cases, disrupted circadian rhythms are associated with slowed metabolism and increased body weight.

Jet Lag

As noted in Chapter 1, jet lag is a relatively common phenomenon experienced by travelers who cross multiple time zones, resulting in shifted light-dark cycles, as well as disrupted meal and sleep patterns. Internal rhythms of hormone secretion, body temperature regulation, and sleep drift away from optimal. Certainly, one factor that may contribute to jet lag is the interaction among circadian rhythms, light exposure, sleep, time of food intake, and metabolism.

For example, an experimental study of jet lag determined that if meals are consumed during the typical "rest" phase, as assessed by elevated melatonin concentrations (melatonin is only secreted during dark nights [Box 2.1]), then decreased insulin sensitivity and impaired metabolism are observed.[8] On the other hand, consuming meals during the "active" phase increases insulin sensitivity and enhances metabolism.[8]

Box 2.1 Melatonin

Melatonin is an ancient hormone, and is present in species from invertebrates to humans. In all animals studied, melatonin is secreted only at night, regardless of whether the individual is diurnal or nocturnal in its habits. Thus, a cell sitting in the liver "knows" that it is nighttime if it detects melatonin. Melatonin is often called the darkness hormone. In mammals such as us, melatonin is produced in the pineal gland. Among humans, as in other mammals, shifting the timing of the light-dark cycle results in a comparable shift in the timing of the nightly peak of melatonin secretion. In a variety of mammals, including humans, the nocturnal synthesis and secretion of pineal melatonin can be rapidly inhibited by exposure to brief periods of light at night. Thus light has two actions in humans, as it does in other mammals: (1) light can entrain, or synchronize, the daily melatonin rhythm, and (2) light can acutely suppress daily melatonin secretion.

We can use the onset of melatonin secretion, which occurs about 14 hours after awakening, as a marker of circadian phase for synchronization of other circadian rhythms, and it may also influence circadian phase. For example, men and women participating in an experiment were housed under dim light and given melatonin at specific times of the day. Under these conditions, melatonin administered for four consecutive days during the late afternoon or early evening tended to advance the onset of internal melatonin secretion, whereas

continued

Box 2.1 *continued*

melatonin given in the morning tended to delay the onset of melatonin secretion.

Knowing these temporal parameters allows melatonin (and light) to be used therapeutically to phase shift rhythms. Thus, light and melatonin can be given to treat ailments such as jet lag, problems associated with night-shift work, advanced and delayed sleep phase disorders, and seasonal affective disorder (SAD).

Another study explored the effects of a high-fat meal on metabolism in individuals experiencing or not experiencing experimental jet lag.[9] Individuals with jet lag displayed slower metabolic rates and lower resting energy expenditures compared to those not experiencing jet lag.[9] These results suggest that jet lag may disrupt the body's ability to effectively metabolize food, potentially contributing to weight gain and associated negative health outcomes.

Overall, it seems that the timing of food intake can significantly influence metabolism, and jet lag may significantly disrupt this process. It is important for travelers to consider the timing of their exposure to light (it may be delayed or advanced several hours) leading to poor alignment of internal rhythms. Also, the timing of meals will be out-of-phase with your metabolism for several days. Try to align light exposure and meal times with your body's natural circadian rhythm to optimize metabolism and potentially alleviate some of the negative effects of jet lag.

Social Jet Lag

Virtually all of us in the modern world suffer regularly from social jet lag, which again is the misalignment of individuals' circadian rhythms with the external social environment.[10] Social jet lag seems to affect college students the most, as they may shift by as many as 6 hours as they stay up late on Friday and Saturday nights and sleep in on weekend mornings. At the end of the weekend people are forced to try to resume their weeknight bedtimes, resulting in difficult Monday mornings. Again, the exposure to light is inappropriately adjusted every few days so that our circadian rhythms remain poorly synchronized with a 24-hour day and our bodily functions are not optimal. In any case, one factor that can contribute to social jet lag is the timing of food intake, as the body's metabolism is influenced by when we eat.[11] Indeed, people experiencing social jet lag of only 1-hour shifts between weekdays and weekends are 1.75 times more likely to be prediabetic or diabetic than those not experiencing social jet lag![12]

Generally, individuals who engage in social jet lag are more likely to have unhealthy eating habits, including late night snacking and irregular mealtimes.[12] These social behaviors can disrupt the body's natural circadian rhythm and lead to weight gain and other metabolic issues, such as insulin resistance.[13,14] If most folks are experiencing social jet lag, then it is not surprising that so many people are gaining weight, and that rates of obesity and diabetes continue to increase in North America, as well as in Europe.

The interaction among social jet lag, light exposure, time of food intake, and metabolism is complex and multifaceted. Whereas social jet lag can lead to unhealthy eating habits and the subsequent negative effects on metabolism, maintaining consistent mealtimes and avoiding late-night eating may help to mitigate these effects.

Exposure to Dim Lighting During the Day

Many of us spend our days in dimly lit homes, offices, factories, hospitals, or assisted living facilities. Not surprisingly, the lack of bright days also influences our daily rhythms, typically not in a good way. That is, exposure to dim illumination during the day also perturbs circadian rhythms and has been linked to increased body mass and altered metabolism; dim daytime lighting can reduce metabolic rate, increase appetite, and increase fat storage, leading to weight gain.[15,16]

In general, dim daytime lighting reduces melatonin production, inhibits the breakdown of fat, and increases hunger and cravings. These changes can lead to increased risk of obesity and metabolic diseases, including type 2 diabetes. Several recent studies have examined the effects of dim light on metabolism, body mass, and other health outcomes. One study reported that people exposed to dim lighting during the day displayed increased BMI, waist circumference, and fat mass. The authors of this study also reported increased concentrations of the "hunger hormone," ghrelin, in people exposed to dim light during the day compared to those exposed to typical lighting.[17]

In a study of people living in rural Brazil, individuals were monitored with wearable devices that recorded locomotor activities and light exposure throughout the day. There were two experimental groups in this study: One group had metabolic syndrome, whereas the other group did not. Motor activity and circadian profiles did not differ between the two groups by the end of the study. However, the individuals with metabolic syndrome displayed low light exposure during the day and high light exposure at night compared to the individuals without metabolic syndrome. The study authors' analyses reported that high daytime light exposure and low light exposure during the night were associated with reduced risk for metabolic syndrome.[18]

In another rigorously designed randomized clinical trial, overweight, insulin-resistant volunteers were exposed to two 40-hour laboratory sessions with different 24-hour lighting protocols. All other variables were controlled, and the participants experienced both sets of experimental conditions. Thus, this was a randomized, controlled, crossover trial. Study participants were exposed either to bright illumination during the day (~1250 lux) and dim light at night (~5 lux) or to dim light during the day and bright light at night. After some time, participants switched lighting conditions, thereby acting as their own experimental controls. This experimental design ensured the most rigorous, unbiased outcomes. During the study participants stayed in a specially equipped chamber called a metabolic chamber that was able to continuously assess their metabolism. There were no changes in sleep among people in the two experimental conditions. However, exposure to bright light during the day lowered glucose levels after dinner (6:00 p.m./1800h), compared to exposure to dim light during the day. Metabolic rate was also reduced under these conditions. Taken together, these results suggest a perfect storm for developing metabolic syndrome in folks exposed to dimly illuminated days or bright light at night conditions by provoking weight gain and increasing post-meal glucose levels.

Late Night Food Intake

During the past 20 years, much research has focused on temporal meal patterns and how the timing of meals affects metabolism. Significant research and clinical management suggest that manipulation of meal timing may be an effective intervention to mitigate the risks of obesity and diabetes that are associated with disrupted circadian rhythms. Indeed, this work forms the basis of intermittent fasting diet programs.[19,20]

As a reminder, intermittent fasting diets focus on the timing of food intake. Unlike most diets that focus on what foods to eat (or more commonly what foods not to eat), intermittent fasting is about when you eat. Under these programs, you typically only eat during a specific time—for example, within an 8-hour window each day—and fast for the remaining 16 hours. Alternatively, some intermittent fasts involved just one meal per day for a couple days each week. This pattern of food intake is consistent with our Paleo ancestors' food availability given the limits of daytime lighting.

When I was a kid, TVs stopped broadcasting after the evening news in the United States. There were no computers, cell phones, or electronic tablets to entertain us. People did not stay up snacking while scrolling through social media or watch TV or movies on their devices. People played and worked outdoors, and most mealtimes ended by 6:00 or 7:00 p.m. Obesity was rare. However, with the light exposure afforded by the Internet, gaming, and social media, combined with ingesting calorie-dense snacks throughout much of the night, obesity rates have soared. The glucose in our blood remains high for much longer than our Paleo (or 20th century) ancestors experienced and is easily converted to fat if not being burned by exercise. By limiting caloric intake to 8 or so hours per day, after a few hours without food, the glucose in the blood is exhausted and the body starts to burn fat during the evening. This intermittent fasting strategy is also effective at controlling blood glucose for people with prediabetes or diabetes.

An extreme example of late-night eaters is individuals with so-called night eating syndrome; a disorder that was first described in the 1950s.[21,22] Individuals with night eating syndrome consume more than 25% of their daily calories after dinner. They often awaken from sleep to eat. Remarkably, individuals with night eating syndrome usually maintain otherwise typical sleep and activity rhythms. Folks who have night eating syndrome are likely to be overweight/obese and are at high risk of developing diabetes. It is interesting to note that nearly half of the people who are extremely obese and selected to undergo bariatric surgery as treatment for their morbid obesity have symptoms that are consistent with the diagnostic criteria associated with night eating syndrome.[22] Individuals diagnosed with night eating syndrome can often be treated successfully with cognitive behavioral therapy.[23] For example, one case study of a 30-year-old male patient with night eating syndrome used several prevention strategies to limit his night-time eating. These strategies included placing signs on his refrigerator, placing a barrier in the hallway between his bedroom and kitchen to awaken him fully, setting a rule to discontinue eating while standing next to the refrigerator, and ultimately locking his refrigerator with a chain.[23]

People with night eating syndrome often display altered hormone concentrations. For example, individuals with night eating syndrome unsurprisingly present with low melatonin concentrations, no doubt reflecting their inopportune exposure to light during late-night snacking. Treatment of individuals with night eating syndrome with timed melatonin treatment reduces both night-time eating behavior and body weight.[21] The efficacy of melatonin in treating night eating syndrome suggests that circadian rhythm disruption may be an important factor in this behavioral disorder. Another contributor to night eating syndrome may be reduced leptin, a satiety hormone. Leptin is typically released by fat cells and serves as a "gauge" of available energy reserves,[24] which is crucial for long-term weight maintenance. Low leptin levels typically occur among individuals with low energy reserves, which has the adaptive effect of increasing the drive to eat. In contrast, when low leptin concentrations occur among individuals with high energy reserves, this can drive maladaptive, excess food consumption that further increases obesity.

Although the consequences are less extreme than night eating syndrome, many individuals may be unwittingly contributing to weight gain simply based on how they distribute their calorie intake across the day. Individuals who skip breakfast and/or consume more of their calories later in the day are more likely to be overweight [25]. For example, in a study of the effects of mealtimes on health, nearly 2,000 Italian men and women documented when and what they ate, and researchers recorded their health parameters over 3 days. Six years later, the researchers re-examined them and discovered that people who ate the majority of high-calorie foods at dinner had increased risk of obesity, metabolic syndrome, and nonalcoholic fatty liver disease.[26] A recent meta-analysis (examination of data from several independent studies of the same topic in order to determine overall trends) of several studies on evening eating habits and obesity in many different countries and cultures also reported a strong link between eating at night and obesity.[27] Finally, analyses of the timing of food intake were conducted using data from the University of California, Los Angeles, Energetics Study.[28] Obesity was associated with mealtime in participants who consumed >33% of their energy foods at dinner, but not among those who consumed >33% of their daily energy food at breakfast or lunch. Later meals are associated with a decrease in resting energy expenditure and a decrease in lipid oxidation without affecting an individual's activity levels.[29] Late or evening meals are also associated with glucose dysregulation, increased fasting glucose, and increased Hb1c (a blood test that provides a lengthened snapshot of blood glucose levels).[29] Taken together, eating more of your daily calories at night results in the development of metabolic syndrome compared to consuming the same number of calories during the day.

Whereas consuming meals at the "wrong" time of day is associated with metabolic impairments, restricting meals to the "correct" time of day has beneficial effects on health. A recent meta-analysis of 19 clinical studies reported that restricting meals to optimal circadian times reduced body weight and fat mass in overweight or obese study participants. Moreover, "correctly" timed meals lowered blood glucose, improved circulating triglyceride levels, and improved systolic blood pressure.[30] Although the study focused on older, unhealthy obese people, the results of healthy young individuals on time-restricted diets did not reveal any significant weight loss, but did show reduced fat mass and increased muscle mass. These results suggest that time-restricted food intake has beneficial effects in people who have and do not have metabolic dysfunction.[30]

The timing of macronutrient content can also impact metabolism. For example, shifting a carbohydrate-heavy meal to the morning (as opposed to the evening) reduces the risk of obesity and metabolic syndrome.[31] However, maintaining a healthy diet, exercise routine, and meal schedule is likely difficult for the average individual, especially when combined with the challenge of disrupted circadian rhythms. Therefore, as growing evidence supports the benefits of correctly timed meals, an emerging question that has yet to be fully addressed is how can one achieve and maintain the timed eating behaviors that are necessary for optimal metabolic health?

Exposure to Light at Night

In addition to our mostly processed-food diet, one feature of modern life that may have even more negative consequences for our metabolism is artificial light at night. Outside of the very highest latitudes, life on Earth evolved over the course of the past 3 to 4 billion years under the defining pattern of bright light present only during the solar day and dark during the night. As noted in Chapter 1, the temporal rhythm of our rotating planet was internalized in our bodies. In common with all other animals on the planet, virtually every aspect of our physiology and behavior, ranging from sleep to hormone secretion, to body temperature regulation, to metabolism and food intake, is regulated by our internal clocks. The increase in exposure to light at night parallels the global increase in the prevalence of obesity and metabolic disorders (Figure 2.1).

As we read earlier in this chapter, our circadian rhythms are primed to be synchronized to precisely 24 hours each day by exposure to bright light during the day and dark environments during the night. However, we sit in dimly lit homes, dormitories, nursing facilities, or offices, and are exposed to significant artificial light at night, and as described, exposure to dim light during the day negatively affects metabolism and promotes weight gain. However, exposure to artificial light at night also disrupts our circadian rhythms.

Exposure to light at night is associated with increased prevalence of obesity in humans.[31] Similarly, there is a positive correlation between the timing of food intake and success in weight reduction; constraining food intake between morning and early evening (e.g., 8:00 a.m. to 4:00 p.m. [0800h–1600h]) increases body weight loss for a number of weight loss diets.[32]

A large correlational study based on data from the National Health and Nutrition Examination Surveys (known as NHANES) indicated that each 3-hour increase in the duration of nighttime fasting was associated with significantly reduced odds of elevated HbA1c.[33] The term HbA1c refers to glycated hemoglobin. It forms when hemoglobin, a protein within the red blood cells that carries oxygen throughout your body, joins with glucose in the blood, becoming "glycated." In contrast to a single test of

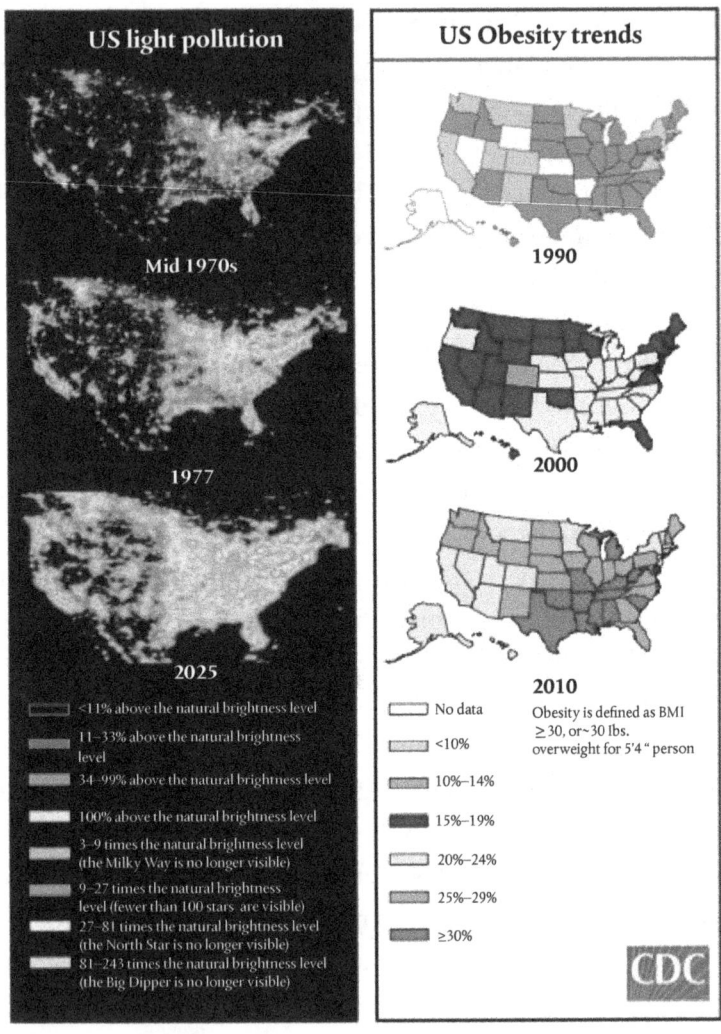

Figure 2.1 Exposure to artificial light at night and rates of obesity are increasing in tandem. The left panel depicts light pollution trends in the United States from the 1970s and projected to 2025. The panel on the right shows US obesity trends from 1990. Images from NASA and the CDC. Reproduced with permission from Fonken L.K., Nelson R.J. (2014). The effects of light at night on circadian clocks and metabolism. Endocrine Reviews, 35: 648–670.

blood glucose levels, which only provide this value at the specific time when tested, these glycated substances serve as a long-term timestamp of your blood glucose values. Thus, by measuring glycated hemoglobin (HbA1c), your healthcare providers get an overall picture of what your average blood sugar levels have been over a period of 1–2 months. For everyone, but especially for people with diabetes, this is important, as the higher the HbA1c values, the greater the risk of developing diabetes-related complications. Indeed, women who consumed fewer than 30% of their daily calories after 5:00 p.m. displayed significantly decreased C-reactive protein (CRP—a blood marker of inflammation) concentrations.[34] Elevated CRP level is a risk factor for diabetes; individuals with prediabetes have higher CRP levels than those with typical glucose concentrations. Importantly, systemic inflammation may play a role in the early-phase deterioration of glucose metabolism at the onset of diabetes, so low CRP levels are important for good overall health.

Of course, the results from human studies are typically correlational and do not establish causation. A correlation is simply a relationship, and one scientific mantra to keep in mind is that "correlation is not causation." For example, you would likely see a strong relationship or correlation between the levels of water in the storm sewer and the number of umbrellas raised in a city. An absence of water in the storm drains would likely be related to observing no umbrellas raised. Several inches of water in the sewer would likely be correlated to many raised umbrellas. Thus, there would probably be a strong correlation between the levels of water in the storm system and number of umbrellas raised; however, we would not say that raising umbrellas causes the level of water to rise in the sewer. In other words, the number of umbrellas and amount of water in the sewer are not directly related to one another. In fact, the causal factor responsible for both the number of raised umbrellas and the amount of water in the sewer is the amount of rain falling from the sky. Nonetheless, the correlational studies in humans are limited, but generally support the hypothesis that consuming energy earlier in the day and prolonging the length of the nightly fast may reduce the risk of several common chronic metabolic disorders. To establish causation, studies of laboratory rodents are often employed.

There have been dozens of studies in rodents showing that the light at night disrupts circadian rhythms, changes the timing of food intake, and promotes elevated blood glucose levels. Many of the effects of light at night on metabolism have been attributed to sleep disruption in people. One important advantage to examining nocturnal mice and rats is that they sleep during the day; light at night disrupts circadian rhythms usually without affecting sleep. Thus, the effects of sleep and disrupted circadian rhythms can be separated in these studies. To summarize these rodent studies, in the absence of disrupted sleep, light at night increases the rate of conversion of food to fat, increases overall body weight, elevates circulating glucose levels, and impairs glucose clearance from the blood. These latter effects are hallmarks of prediabetes and type 2 diabetes. These effects occur despite the animals' expending the same number of calories and consuming the same number of calories each day!

Depending on the timing of the light at night, the major mechanisms contributing to altered glucose homeostasis differ. Light during the early dark phase increases circulating glucose levels likely via a reduction in a glucose transporter molecule (GLUT4; responsible for ferrying glucose into muscle cells) expression in muscle,[35] whereas light during the late dark phase is associated with a reduction in insulin sensitivity.[36] Insulin insensitivity (or insulin resistance) can occur in response to several reasons that cause your muscle, fat, and liver cells to respond inappropriately to insulin. In response to insulin, glucose typically enters your cells and the levels of glucose in your bloodstream decrease; this reduction in circulating glucose signals your pancreas to stop producing insulin. As your pancreas makes more insulin to try to overcome your increasing blood glucose levels, a condition called hyperinsulinemia ensues. This can be temporary or more chronic—chronic hyperinsulinemia can exhaust insulin production, and when this happens, people must provide insulin from external sources to support their metabolism.

The metabolic effects of light at night in rodents seem to reflect a shift in the timing of their food consumption. Although the number of calories consumed and the calories expended each day do not change, exposure to light at night shifted the time of food intake to the animals' subjective resting period (i.e., the "wrong" time to eat). If food access was restricted to the appropriate time of day (the active period), then animals exposed to light at night did not display increased body mass or changes in glucose metabolism. Providing appropriately timed (at the onset of the night) melatonin treatment during exposure to light at night improves metabolic outcomes by mitigating disrupted circadian rhythms. Importantly, provisioning mice with high fat diets and then exposing them to light at night increased the body weight increase, and also provoked diabetes![37,38] Combining melatonin with metformin, a common diabetes and prediabetes drug, synergistically improved glucose regulation more than either treatment alone.[39] Although research on treatment options is ongoing, it is likely that a combination of approaches that target both the underlying disruption of circadian rhythms and the metabolic impairments will yield the most promising results in reversing the obesity and diabetes associated with artificial light at night.[40]

Night-Shift Work

Night-shift work is the trifecta of disrupted circadian rhythms— night-shift workers are exposed to light at night, they eat at the "wrong time" of day, and they are often exposed to frequent phase shifts as they move between day and night shifts or between their work schedules and weekend schedules with friends and family. In a sense, night-shift workers, like my friend Jennifer, who are exposed to high levels of light at night, mistimed meals, and recurring phase shifts have served as a warning to the rest of us of the maladaptive consequences of disruptive circadian rhythms. There is a large and robust scientific literature suggesting night-shift work is associated with nighttime eating and increased risks of obesity and metabolic diseases such as diabetes and cardiovascular disorders compared to its day-shift counterpart.

I worked night shifts at a local turkey abattoir during high school. Night-shift workers are awake, active, eating, and drinking when our circadian rhythms are designed to compel us to rest (i.e., night). I typically worked night shifts from Sunday to Thursday nights. Then, I would usually adopt a more typical schedule on the weekends so that I could hang out with my friends, shifting my activities, mealtimes, sleep times, and light exposure schedules by 8–12 hours.

Working night shifts provoked a profound mismatch between my physiology and behavior and the external solar day. Such a mismatch disrupts the internal circadian rhythms on the molecular, cellular, tissue, and behavior levels. Only about 2% of night-shift workers fully embrace night-shift schedules; in other words, only 2% remain active and awake on their nights off work to sustain a consistent internal schedule.[41] After graduate school, I was able to shift my routine to one more aligned with the solar day, but about 20% of us in the United States or Europe currently work the night shifts. Night-shift workers are common in healthcare, first responders, manufacturing, transportation, and many aspects of retail.

As mentioned, night-shift workers are at increased risk for several cardiometabolic diseases as compared to people working comparable jobs during the day shift. The increased incidence of obesity and diabetes associated with working the night shift suggests that disrupted circadian rhythms promote increased body mass and glucose dysregulation. Approximately half of night-shift workers in the United States are overweight or obese; working night shifts has been estimated to increase type 2 diabetes by nearly 10%![42]

For example, more offshore oil-rig workers who go to work on North Sea oil platforms become obese than those who stay local. It was initially thought that this reflected a difference in oil-rig workers eating high-calorie foods and the lack of physical exercise.[43] However, additional follow-up studies indicated that the amount of time spent working on night shifts was most closely associated with increases in body weight among the oil-rig workers.

Other researchers reported similar relationships between the night shift in other settings and metabolic disorders. For instance, male night-shift workers at a Japanese steel mill, as well as night-shift workers at a zipper factory in Japan, were more likely to be obese than day-shift workers at the same job.[44,45]

We have learned much about the role of night-shift work by studying nurses like my friend Jennifer. In accordance with the previously described studies, current night-shift work is associated with obesity in studies of Brazilian and Polish nurses.[46,47] Remarkably, the years spent working night shifts as a nurse are also associated with obesity. In the study of Polish nurses, those who worked eight or more night shifts per month during their career retained the likelihood of becoming obese.[47] That is, the effects of night-shift work on metabolism seem to persist because former night-shift workers retained a higher risk of obesity even after they switched to day shifts!

Perhaps more sobering is the observation that some studies have reported an association between night-shift work and other metabolic risks. For example, a positive relationship between working night shifts and an elevated risk for type 2 diabetes has been reported as part of the Nurses' Health Studies. Nurses were tracked in two groups. In the first group, about 108,000 nurses aged 25–47 years were monitored from 1989 to 2007. In the second group, nearly 70,000 nurses aged 42–67 years were monitored from 1998 to 2008. It was determined not only that women who worked night shifts had a higher risk for type 2 diabetes but also that the risk was proportionally linked to how many years they worked night shifts.[48]

As mentioned, what likely makes night-shift work so detrimental to health is that night-shift individuals experience many factors working against their circadian rhythms. Night-shift workers experience increased exposure to light, elevated caloric intake, and reduced sleep during their "night," which usually is attempted in a relatively bright bedroom illuminated by the sun. Additional bad news for night-shift workers is that they tend to eat high-calorie foods and display low levels of activity relative to individuals who perform the same jobs during the day shift.[49,50] Although few studies have been reported, night-shift workers who minimize the effects of artificial light at night by using blue light-blocking glasses may potentially decrease their risk of obesity and diabetes.

Again, feeding signals appear to be the dominant timing cue for the rhythms of peripheral clocks in the liver, pancreas, and gut, including those that control metabolic pathways. Thus, consuming food outside of the typical time of eating (i.e., late-night eating in humans) appears to reset some peripheral clocks and disrupt energy balance. Being mindful of when and what types of food are eaten while working the night shifts is important in combating the metabolic challenges.

To mimic night shifts in mouse models of disrupted circadian rhythms, exposure to light at night is often used as described. However, changing the light-dark cycles has been one of the most common means of disrupting circadian rhythms. Such adjustments to the environmental lighting cycles include models of jetlag (e.g., 1- to 8-hour phase shifts—turning on or off the lights at different times), light-dark inversions (e.g., a 12-hour phase shift for animals maintained on 12 hours of light and 12 hours of dark day [LD 12:12]), or noncompatible, unsynchronized 24-hour days (e.g., 20- or 28-hour days).

Typically, minor shifts in the light-dark cycle in isolation are insufficient to provoke measurable effects in studies of rodent metabolism. For example, the short-term housing of mice on non-24-hour days does not affect their body weight or glucose tolerance.[51] Of course, it is probably to be expected that such experimental conditions do not completely mimic the metabolic effects reported for night-shift workers. Rarely do animal studies that change the light-dark cycles include the presence of light at night, altered changes in patterns of eating or sleeping, the freedom to select different food types (e.g., high fat or lean) over the day, or social jetlag. The metabolic effects of disrupted rodent circadian rhythms begin to reflect the observations in humans when some of these conditions are included.

One interesting study allowed rats free access to their typical food during the entire day, but provided a chocolate "treat" at different times of the day in combination with various shifts in the light-dark cycles. Chocolate served as a model for the high fat/sugar food favored by night-shift workers. Using this model, rats given the chocolate at the "wrong" time of day experienced maximum disruption of their circadian rhythms and also displayed maximum gains in body weight.[52]

Rodent studies provide stronger and more direct evidence that, in a remarkably short time, disrupted circadian rhythms shift the timing of food intake with subsequent metabolic problems, including elevated blood glucose and insulin concentrations, and even type 2 diabetes!

Optimal Time for Food Intake

If you are interested in reducing or maintaining your body weight, then when is the best time to eat? In a recent 12-week study,[53] 50 overweight women were randomly assigned to a 1,400-calorie diet that comprised breakfast of 700 calories, lunch of 500 calories, and dinner of 200 calories, or they received the same calories and food choices, but with the breakfast (200 calories) and dinner (700 calories) meals switched. Although women in both groups lost significant amounts of weight, the women consuming their highest caloric meal at breakfast lost an average of ~19 pounds (~8.6 kg) compared to women eating their highest caloric meal at dinner, who lost only about 8 pounds (3.6 kg) during the same time. The large breakfast group also lost twice as many inches (cm) around their waists compared to the women eating large dinners.

Also, many epidemiological studies have indicated a direct relationship between obesity and light at night. For example, in one study of 100,000 women in the United Kingdom, the odds of obesity as assessed by BMI, waist–hip ratio, and waist circumference increased with elevated exposure to light at night.[54] This correlation was related only to exposure to light at night and was not related to sleep duration, alcohol intake, cigarette smoking, or physical activity.

Several studies on rodents indicate that light at night suppresses the nocturnal rise in melatonin concentrations. Restoration of melatonin rhythms appears to improve blood glucose and body weight regulation,[55] whereas reduced night-time secretion of melatonin is associated with an increased risk of developing type 2 diabetes in humans.[56] The temporal pattern of light exposure also appears important for weight regulation. Again, in a 10-day inpatient study of disrupted circadian rhythms, several metabolic hormones, including leptin, glucose, and insulin were altered to the extent that by the end of the study, some of the healthy volunteers showed glucose levels consistent with prediabetes.[7] Taken together, disruption of circadian rhythms evokes several lifestyle and behavioral changes that may further affect metabolism, but most notably, disruption of circadian organization can result in a mismatch between the biological clock and timing of food intake.

Thus, to improve metabolism and maintenance of a healthy body weight, try to resynchronize your circadian rhythms. Avoid light at night, but seek out bright light during the day. Avoid late night food intake. Fast between dinner and breakfast, preferably 14 hours. The same holds true for night-shift workers. It is hypothesized that fasting and time-restricted feeding regimens that actively impose a diurnal rhythm of food intake that is aligned with the 24-hour solar day lead to improved daily cycles in circadian clock gene expression, reprogramming of molecular mechanisms of energy metabolism, improved insulin metabolism, and improved body weight regulation.[57,58] It is important for night-shift workers to be aware of the potential negative impacts on their health and make an effort to maintain a healthy diet and regular mealtimes. This may involve bringing healthy snacks to work or planning meals in advance to ensure that they are not relying on fast food or vending machine options. This may also involve blocking out short wavelength (blue) light with special blue-light-blocking goggles. Again, it is critical for night-shift workers to be especially mindful of their diet and make an effort to maintain healthy eating habits.[59,60]

References

1. Colin J, Timbal J, Boutelier C, Houdas Y, Siffre M. 1968. Rhythm of the rectal temperature during a 6-month free-running experiment. *Journal of Applied Physiology*, 25: 170–176.

2. Halberg F, Reinberg A, Haus E, Ghata J, Siffre M. 1970. Human biological rhythms during and after several months of isolation underground in natural caves. *Bulletin of the National Spelunking. Society*, 32: 89–115.

3. Nelson RJ, Chbeir S. 2018. Dark matters: effects of light at night on metabolism. *Proceedings of the Nutrition Society*, 77: 223–229.

4. Ekirch AR. 2013. *At Day's Close: A History of Nighttime*. W.W. Norton, New York.

5. Moran-Ramos S, Baez-Ruiz A, Buijs RM, Escobar C. 2016. When to eat? The influence of circadian rhythms on metabolic health: are animal studies providing the evidence? *Nutrition Research Reviews*, 29: 180–193.

6. Pattinson CL, Allan AC, Staton SL, Thorpe KJ, Smith SS. 2016. Environmental light exposure is associated with increased body mass in children. *PLoS One*, 11: e0143578.

7. Wefers J, van Moorsel D, Hansen J, Connell NJ, Havekes B, Hoeks J, van Marken Lichtenbelt WD, Duez H, Phielix E, Kalsbeek A, Boekschoten MV, Hooiveld GJ, Hesselink MKC, Kersten S, Staels B, Scheer, Schrauwen P. 2018. Circadian misalignment induces fatty acid metabolism gene profiles and compromises insulin sensitivity in human skeletal muscle. *Proceedings of the National Academies of Sciences (USA)*, 115: 7789–7794.

8. Okamoto K, Fujimoto K, Goto M, et al. 2013. Effect of jet lag on energy metabolism in humans. *Journal of Physiology*, 591: 1655–1666.

9. Klerman EB, Dijk DJ, Rusterholz T. 2018. Social jetlag, chronotype and human health. *Current Opinion in Psychology*, 21: 45–50.

10. Farshchi HR, Taylor MA, Macdonald IA. 2005. Beneficial metabolic effects of regular meal frequency on dietary thermogenesis, insulin

sensitivity, and fasting lipid profiles in healthy obese women. *American Journal of Clinical Nutrition*, 81: 16–24.

11. Koopman ADM, Rauh SP, van 't Riet E, Groeneveld L, van der Heijden AA, Elders PJ, Dekker JM, Nijpels G, Beulens JW, Rutters F. 2017. The association between social jetlag, the metabolic syndrome, and type 2 diabetes mellitus in the general population: the New Hoorn Study. *Journal of Biological Rhythms*, 32: 359–368.

12. Knutson KL, Spiegel K, Penev P, Van Cauter E. 2007. The metabolic consequences of sleep deprivation. *Sleep Medicine Reviews*, 11: 163–178.

13. Scheer FAJL, Hilton MF, Mantzoros CS, Shea SA. 2009. Adverse metabolic and cardiovascular consequences of circadian misalignment. *Proceedings of the National Academies of Sciences (USA)*, 106: 4453–4458.

14. Klerman EB, Dijk DJ, Rusterholz T. 2018. Social jetlag, chronotype and human health. *Current Opinions in Psychology*, 21: 45–50.

15. Drouot L, Touitou Y. 2017. Effects of dim light at night on circadian clocks and metabolism. *Chronobiology International*, 34: 1040–1048.

16. Oren E, Schuldiner S. 2018. The effect of dim light during the day on circadian timing and metabolic functions. *Chronobiology International*, 35: 990–996.

17. Arble DM. 2023. Disrupted circadian rhythms and metabolic function. In: *Biological Implications of Circadian Disruption: A Modern Health Challenge*, LK Fonken, RJ Nelson (Eds). Cambridge University Press, Cambridge, UK.

18. Benedito-Silva AA, Evans S, Viana Mendes J, Castro J, Gonçalves BDSB, Ruiz FS, Beijamini F, Evangelista FS, Vallada H, Krieger JE, von Schantz M, Pereira AC, Pedrazzoli M. 2020. Association between light exposure and metabolic syndrome in a rural Brazilian town. *PLoS One*, 15: e0238772.

19. Panda S. 2018. *The Circadian Code: Lose Weight, Supercharge Your Energy, and Transform Your Health from Morning to Midnight*. Rodale Books, New York.

20. Panda S. 2021. *The Circadian Diabetes Code*. Rodale Books, New York.

21. Pinto TF, Silva FG, Bruin VM, Bruin PF. 2016. Night eating syndrome: how to treat it? *Revista da Associação Médica Brasileira (1992)*, 62: 701–707.

22. Stunkard AJ, Grace WJ, Wolff HG. 1955. The night-eating syndrome: a pattern of food intake among certain obese patients. *American Journal of Medicine*, 19: 78–86.

23. Berner LA, Allison KC. 2013. Behavioral management of night eating disorders. *Psychological Research and Behavioral Management*, 6: 1–8.

24. Kelesidis T, Kelesidis I, Chou S, Mantzoros CS. 2010. The role of leptin in human physiology: emerging clinical applications. *Annals of Internal Medicine*, 152: 93–100.

25. Singh RB, Cornelissen G, Mojto V, et al. 2020. Effects of circadian restricted feeding on parameters of metabolic syndrome among healthy subjects. *Chronobiology International*, 37: 395–402.

26. Bo S, Musso G, Beccuti G, Fadda M, Fedele D, Gambino R, Gentile L, Durazzo M, Ghigo E, Cassader M. 2014. Consuming more of daily caloric intake at dinner predisposes to obesity: a 6-year population-based prospective cohort study. *PLoS One*, 9: e108467.

27. Almoosawi S, Vingeliene S, Karagounis LG, et al. 2016. Chrono-nutrition: a review of current evidence from observational studies on global trends in time-of-day of energy intake and its association with obesity. *Proceedings of the Nutrition Society*, 75: 487–500.

28. Wang JB, Patterson RE, Ang A, Emond JA, Shetty N, Arab L. 2014. Timing of energy intake during the day is associated with the risk of obesity in adults. *Journal of Human Nutrition and Diet*, 27: 255–262.

29. Kelly KP, McGuinness OP, Buchowski M, Hughey JJ, Chen H, Powers J, Page T, Johnson CH. 2020. Eating breakfast and avoiding late-evening snacking sustains lipid oxidation. *PLoS Biol*, 18: e3000622.

30. Moon S, Kang J, Kim SH, Chung HS, Kim YJ, Yu JM, Cho ST, Oh C-M, Kim T. 2020. Beneficial effects of time-restricted eating on metabolic diseases: a systemic review and meta-analysis. *Nutrients*, 12: 1267.

31. Almoosawi S, Prynne CJ, Hardy R, Stephen AM. 2013. Time-of-day and nutrient composition of eating occasions: prospective association with the metabolic syndrome in the 1946 British birth cohort. *International Journal of Obesity (London)*, 37: 725–731.

32. Obayashi K, Saeki K, Iwamoto J, Okamoto N, Tomioka K, Nezu S, Ikada Y, Kurumatani N. 2013. Exposure to light at night, nocturnal urinary melatonin excretion, and obesity/dyslipidemia in the elderly: a cross-sectional analysis of the HEIJO-KYO study. *Journal of Clinical Endocrinology and Metabolism*, 98: 337–344.

33. Marinac CR, Natarajan L, Sears DD, Gallo LC, Hartman SJ, Arredondo E, Patterson RE. 2015. Prolonged nightly fasting and breast cancer risk: findings from NHANES (2009–2010). *Cancer Epidemiological Biomarkers and Prevention*, 24: 783–789.

34. Ginde AA, Cagliero E, Nathan DM, Camargo CA Jr. 2008. Value of risk stratification to increase the predictive validity of HbA1c in screening for undiagnosed diabetes in the US population. *Journal of General and Internal Medicine*, 23: 1346–1353.

35. Nagai N, Ayaki M, Yanagawa T, et al. 2019. Suppression of blue light at night ameliorates metabolic abnormalities by controlling circadian rhythms. *Investigatory Ophthalmology and Vision Science*, 60: 3786–3793.

36. Opperhuizen AL, Stenvers DJ, Jansen RD, Foppen E, Fliers E, Kalsbeek A. 2017. Light at night acutely impairs glucose tolerance in a time-, intensity- and wavelength-dependent manner in rats. *Diabetologia*, 60: 1333–1343.

37. Fonken LK, Lieberman RA, Weil ZM, Nelson RJ. 2013. Dim light at night exaggerates weight gain and inflammation associated with a high-fat diet in male mice, *Endocrinology*, 154: 3817–3825.

38. Fonken LK, Workman JL, Walton JC, Weil ZM, Morris JS, Haim A, Nelson RJ. 2010. Light at night increases body mass by shifting the time of food intake. *Proceedings of the National Academies of Sciences (USA)*, 107: 18664–18669.

39. Thomas AP, Hoang J, Vongbunyong K, Nguyen A, Rakshit K, Matveyenko AV. 2016. Administration of melatonin and metformin prevents deleterious effects of circadian disruption and obesity in male rats. *Endocrinology*, 157: 4720–4731.

40. Arble DM, Sandoval DA, Turek FW, Woods SC, Seeley RJ. 2015. Metabolic effects of bariatric surgery in mouse models of circadian disruption. *International Journal of Obesity (London)*, 39: 1310–1318.

41. Gamble KL, Motsinger-Reif AA, Hida A, Borsetti HM, Servick SV, Ciarleglio CM, Robbins S, Hicks J, Carver K, Hamilton N, Wells N, Summar ML, McMahon DG, Johnson CH. 2011. Shift work in nurses: contribution of phenotypes and genotypes to adaptation. *PLoS One*, 6: e18395.

42. Gan Y, Yang C, Tong X, et al. 2015. Shift work and diabetes mellitus: a meta-analysis of observational studies. *Occupational and Environmental Medicine*, 72: 72–78.

43. Parkes KR. 2003. Demographic and lifestyle predictors of body mass index among offshore oil industry workers: cross-sectional and longitudinal findings. *Occupational Medicine*, 53: 213–221.

44. Suwazono Y, Dochi M, Sakata K, et al. 2008. A longitudinal study on the effect of shift work on weight gain in male Japanese workers. *Obesity*, 16: 1887–1893.

45. Morikawa Y, Nakagawa H, Miura K, et al. 2007. Effect of shift work on body mass index and metabolic parameters. *Scandinavian Journal of Work, Environment and Health*, 33: 45–50.

46. Griep RH, Bastos LS, Fonseca Mendez MJ, Sliva-Costa A, Portela LF, Toivanen S, Rotenberg L. 2014. Years worked at night and body mass index among registered nurses from eighteen public hospitals in Rio de Janeiro, Brazil. *BMC Health Services Research*, 14: 603.

47. Peplonska B, Bukowska A, Sobala W. 2015. Association of rotating night shift work with BMI and abdominal obesity among nurses and midwives. *PLoS One*, 10: e0133761.

48. Pan A, Schernhammer ES, Sun Q, Hu FB. 2011. Rotating night shift work and risk of Type 2 diabetes: two prospective cohort studies in women. *PLoS Med*, 8: e1001141.

49. Bonnell EK, Huggins CE, Huggins CT, McCaffrey TA, Palermo C, Bonham MP. 2017. Influences on dietary choices during day versus night shift in shift workers: a mixed methods study. *Nutrients*, 9: 193.

50. de Assis MA, Nahas MV, Bellisle F, Kupek E. 2003. Meals, snacks and food choices in Brazilian shift workers with high energy expenditure. *Journal of Human Nutrition and Diet*, 16: 283–289.

51. Zitting KM, Vetrivelan R, Yuan RK, Vujovic N, Wang W, Bandaru SS, Quan MB, Williams JS, Duffy JF, Saper CB, Czeisler CA. 2022. Chronic circadian disruption on a high-fat diet impairs glucose tolerance. *Metabolism*, 130: 155158.

52. Escobar C, Espitia-Bautista E, Guzmán-Ruiz MA, Guerrero-Vargas NN, Hernández-Navarrete MÁ, Ángeles-Castellanos M, Morales-Pérez B, Buijs RM. 2020. Chocolate for breakfast prevents circadian desynchrony in experimental models of jet-lag and shift-work. *Scientific Reports*, 10: 6243.

53. Lowe DA, Wu N, Rohdin-Bibby L, Moore AH, Kelly N, Liu YE, Philip E, Vittinghoff E, Heymsfield SB, Olgin JE, Shepherd JA, Weiss EJ. 2020. Effects of time-restricted eating on weight loss and other metabolic parameters in women and men with overweight and obesity: the

TREAT randomized clinical trial. *JAMA Internal Medicine*, 180: 1491–1499.

54. McFadden E, Jones ME, Schoemaker MJ, Ashworth A, Swerdlow AJ. 2014. The relationship between obesity and exposure to light at night: cross-sectional analyses of over 100,000 women in the Breakthrough Generations Study. *American Journal of Epidemiology*, 180: 245–250.

55. Bedrosian TA, Fonken LK, Nelson RJ. 2016. Endocrine effects of circadian disruption. *Annual Review of Physiology*, 78: 109–131.

56. McMullan CJ, Schernhammer ES, Rimm EB, Hu FB, Forman JP. 2013. Melatonin secretion and the incidence of type 2 diabetes. *Journal of the American Medical Association*, 309: 1388–1396.

57. Verdú E, Ramón RM, Farré R. 2017. Meal frequency and weight loss: a systematic review. *Nutrients*, 9: 635.

58. Cao L, Wang P, Chen X, Duan X, Zhu X, Mi M. 2016. Greater proportion of dietary energy intake at breakfast is associated with lower body mass index and body fat in Chinese adults. *American Journal of Clinical Nutrition*, 103: 147–152.

59. Haus E, Smolensky M, Lamberg L. 2000. The role of circadian rhythms in the risk of developing diabetes mellitus. *Diabetes Care*, 23: 360–368.

60. Sasaki H, Ohira T, Miyamoto T, Matsuo T, Kojima T, Nagai K. 2007. Shift work and the risk of metabolic syndrome among Japanese men and women. *Occupational and Environmental Medicine*, 64: 519–524.

3

LIGHT AND MOOD

Light and mood have long been linked figuratively. Light is associated with good, upbeat moods. People are often said to be in sunny or bright moods when they feel positive or uplifted. In contrast, people who may be down or slightly depressed are often considered to be in dark or gloomy moods. This association occurs across many languages and many cultures. For example, In Japanese, the term "明るい" (*akarui*) means bright or cheerful, and it is commonly used to describe a positive mood. On the other hand, "暗い" (*kurai*), meaning dark or gloomy, is used for a negative mood. In Mandarin, "明朗" (*mínglǎng*) means clear and bright and is used to describe a cheerful disposition. Conversely, "阴郁" (*yīnyù*), meaning gloomy or depressed, is used for a darker mood. When feeling "blue" or mildly (not clinically) depressed, several mood elevators are often effective and many of them are associated with getting outdoors during the day; for example, going for a walk or run, gardening, or biking are activities that often help to improve moods in folks who are not diagnosed as clinically depressed.

Dark Matters. Randy J. Nelson, Oxford University Press. © Oxford University Press (2025).
DOI: 10.1093/9780197639979.003.0003

Sleep is critically important for mood, but the extent to which disrupted sleep, as opposed to disrupted circadian rhythms, plays a role remains largely unspecified in people. Sleep is one hand, albeit an important and salient hand, of our biological clock. There is substantial research linking mood to circadian rhythms and sleep. For many mood disorders, a bidirectional relationship exists between impaired circadian rhythms and mood (Figure 3.1). For instance, depressed mood influences the wake-sleep rhythms, circadian rhythms in hormone secretion (especially cortisol and melatonin secretion), and food intake. I will focus on the influence of circadian rhythms on only a few mood disorders, primarily depression, but will describe circadian rhythm influences on additional mood disorders when relevant, as well. The role of disrupted circadian rhythms on mood will be featured throughout the chapter.

The symptoms of clinical depression typically include reduced mood (intense sadness), feelings of worthlessness, feelings of guilt, indecisiveness, disturbed sleep (usually reduced sleep, but

Depressed Mood

Changes Circadian Rhythms of:	Environmental Perturbations
Sleep	Jet Lag
Melatonin	Social Jet Lag
Body Temperature	Night Shift Work
Cortisol Secretion	Short Day Lengths
Metabolism	Daytime Light Deprivation
Clock Gene Expression	Constant Light

Figure 3.1 A reciprocal association between circadian rhythms and mood exists. Depression changes circadian rhythms, and temporal environmental factors can induce depression.

increased sleep is also observed), altered food intake, fatigue, lack of pleasure (anhedonia), suicidal ideation, and restless or blunted motor symptoms.[1] Restless motor symptoms include pacing or handwringing; blunted motor symptoms comprise slow body movements or speech. In severe cases of depression, delusions and hallucinations may occasionally be observed. People displaying symptoms of depression are considered to be suffering from "clinical" depression. If these symptoms persist for more than two weeks, then the individual is typically diagnosed with major depressive disorder (MDD). Adults displaying less severe symptoms lasting at least two years (or one year in children or adolescents) are diagnosed with a relatively mild mood disorder called dysthymia. Another distinction made among types of clinical depression is between bipolar and unipolar depression.

Bipolar depression represents a mood disorder that has two distinct opposite "poles" in mood. Bipolar depression is associated with at least one episode of mania; mania comprises excessively elevated mood, unwarranted feelings of confidence and power, and increased creative energies. The manic phases of bipolar disorder display cycles that may be linked to internal clocks. Some of these cycles may be seasonal. In contrast to bipolar depression, unipolar depression presents as depressed mood in the absence of the manic episodes. Many of the studies on circadian rhythms and depression commonly included people diagnosed with either unipolar or bipolar depression. The physiological mechanisms that contribute to these two types of depression are likely different, but experts differ on the underlying processes and causes. It is important to appreciate that combining of all these individuals into one "depressed" experimental group may mask the importance of the role that circadian rhythms or malfunctioning clock genes play in depression.

Depression is typically considered to be at one end of a mood continuum, with mania at the other end. Mood states between depression and mania include melancholia, typical mood, and hypomania. Depressed individuals vary in both the severity and the duration of their symptoms. The prevalence of depressive symptoms is typically determined via psychological tests such as the Hamilton Rating Scale for Depression, the Hamilton Depression Inventory, and the Beck Depression Inventory.

Major depressive disorder is characterized by persistent low mood and anhedonia. Diagnostic criteria require one or both of these symptoms (i.e., low mood and anhedonia) to be present and accompanied by at least four additional symptoms that include changes in sleep, body weight, and general activity as well as fatigue, feelings of worthlessness, difficulty concentrating, and suicidal thoughts over the same 2-week period; these symptoms should be sufficiently severe to interfere with daily life. MDD is a leading cause of disability affecting nearly 300 million people or 4% of the population worldwide. Importantly, the incidence of MDD worldwide appears to be on the rise. The number of people diagnosed with depression increased by ~18% from 2005 to 2015. Part of this increase may reflect better diagnostic criteria and reduced stigma associated with mood disorders, but across many countries and cultures the increased rates in MDD correlate with modernization of societies.[2] In addition to other environmental factors, as societies modernize, sources of disrupted circadian rhythms (e.g., light at night, night-shift work, and jet lag) proliferate.[3]

The Amish are a protestant sect living primarily in North America. As part of their religious beliefs, they tend to reject modern technology such as automobiles, electric lighting, and all the devices powered by electricity, such as televisions, computers, and cell phones. They use candles and kerosene lanterns for lighting. Amish people experience few disruptions to their circadian rhythms because of their lack of exposure to light at night and social jet lag. Although only a correlation, it is interesting to note that in contrast to other segments of modern society, the rate of MDD has not substantially changed over the years among the Amish.[4] Thus, it seems reasonable to propose that limited exposure to light at night among Amish people may help to explain their stable rates of MDD.

The relationship between the circadian system and MDD is readily apparent on examination of patients' presentation of symptoms. Individuals with MDD often show variation in symptom severity throughout the day with some exhibiting worsening symptoms in the morning, whereas others experience worsening symptoms in the evening. The timing of the most severe symptoms is often associated with different subtypes of MDD. Individuals whose symptoms are worse in the morning display more severe depression, whereas people who display worse symptoms at night are said to display MDD with atypical features. The severity of MDD is also correlated with the extent of misalignment of circadian rhythms;[5] that is, as circadian rhythms become more disorganized, the depressive symptoms worsen. Postmortem examination of brains of patients with MDD reveals impaired patterns of circadian gene expression. Canonical circadian clock genes showed reduced amplitude, shifted peak expression, and alterations in the temporal relationship among genes.[6]

Most successful treatments of depression, including antidepressants (SSRIs, SNRIS, and agomelatine), bright light therapy, wake therapy, and social rhythm therapy, directly influence circadian rhythms.[7] Treatment with morning bright light therapy adjusts the rhythm of melatonin secretion to occur earlier (phase advance) and the extent of phase advancement is correlated with the amelioration of depressive symptoms. Treatment with agomelatine, a drug that binds to melatonin receptors and thereby mimics the biological effects of melatonin (also called a receptor agonist), phase advances circadian rhythms (i.e., adjusts rhythms of cortisol, melatonin, body temperature, and locomotor activity rhythms to be earlier). Indeed, patients diagnosed with depression admitted to hospital rooms facing southeast and thus receiving morning bright light spend half the time in hospital compared to similar patients admitted to northwest-facing rooms.[8,9]

Another indication that MDD is influenced by circadian rhythms is the daily distribution of peak functioning. People can take tests to determine the extent to which they are "night owls" or "morning larks" or fall somewhere in between these extreme states. These so-called chronotypes are officially divided into morningness and eveningness chronotypes (peak alertness early versus peaking later in the day). Some people display extreme chronotypes—that is, either extreme morning types or evening types, but most people are somewhat in between these extremes. There are several free tests online to determine your chronotype. Chronotype is influenced by genetics, age, sex, and patterns of daily light exposure.[10] Chronotypes are not fixed traits that are locked throughout your life. Indeed, chronotype becomes progressively later throughout adolescence and into adulthood, and then adults shift from being night owls back to being morning larks as they age. In any case, depressed individuals tend to show a bimodal distribution of peak function (i.e., best early in the morning and evening with a lull in between), whereas nondepressed people display a unimodal distribution of peak function (i.e., best performance at one time of day). That is, nondepressed people display a single peak of function either during the morning or evening, or somewhere in between.

Anxiety disorders are characterized by intense, excessive, and persistent worry and fear about everyday situations. Rapid pulse rate and breathing, perspiration, and feeling tired may occur. More than 30% of adults in the United States experience anxiety disorders during some period of their lives. Anxiety can be typical in stressful situations such as public speaking or test taking. Anxiety is only an indicator of an underlying mental health disorder when these feelings become excessive, all-consuming, and interfere with daily living.[11] There are five types of primary anxiety disorders:

1. Generalized Anxiety Disorder (GAD). People with GAD tend to feel excessive anxiety or worry on most days for at least six straight months. They may also experience disturbed sleep and exhaustion, and feel restless and "on edge."
2. Panic Disorder. Women are more likely to experience panic disorder than men. People with panic disorder experience panic attacks. Panic attacks are sudden feelings of fear/terror when there is no real danger present. Physical symptoms of panic attacks include rapid heartbeat, chest

or stomach pain, and difficulty breathing. People may also feel weakness or dizziness, sweating, chills, or numbness in their extremities. Many people experiencing panic attacks report that it feels like having a heart attack.

3. Obsessive-Compulsive Disorder (OCD). People with OCD have frequent and often upsetting obsessions (thoughts) or compulsions (coping behaviors). Common obsessions include fear of dirt or germs, fear of getting injured, or a desire to arrange things in certain ways. People with OCD often fixate on these obsessions and can have thoughts and rituals that interfere significantly with their daily life. For example, someone who is worried that their house may catch on fire might check their stovetops multiple times and even be too frightened to leave their house. OCD tends to have a genetic basis and to run in families.

4. Phobia Disorders. People with phobia disorders display intense fear of or aversion to specific objects or situations. This fear is typically far out of proportion to the actual danger caused by the situation or object. People with a phobia may experience irrational worry and take steps to avoid the object or situation. They may also experience immediate anxiety upon encountering the feared object or situation. Common phobias include flying, heights, certain animals (e.g., spiders), and crowds.

5. Post-Traumatic Stress Disorder (PTSD). Individuals suffering from PTSD have experienced a traumatic event such as war, a natural disaster, a serious accident, or physical or sexual abuse. PTSD may cause recurrent flashbacks to the event, trouble sleeping or nightmares, lonely feelings, or angry outbursts. People with PTSD may feel excessively worried,

guilty, or sad. Disrupted circadian rhythms are associated with anxiety, and may contribute to the development of anxiety disorders.[12]

Jet Lag

After crossing multiple time zones, your internal rhythms of hormone secretion, body temperature regulation, and sleep are no longer linked together, but rather drift apart from typical. Thus, you might experience a peak in your body temperature during the middle of the night and wake up in a sweat during the first few days of your European vacation. Formally, jet lag disorder (JLD) is recognized by the American Academy of Sleep Medicine with the following diagnostic criteria[13]:

1. Complaint of insomnia or excessive daytime sleepiness associated with two or more hours of transmeridian (east-west) jet travel.
2. Impaired daytime function, general malaise, or somatic symptoms within 1–2 days of travel.
3. There is no better explanation for the impairment by another sleep, mental, or neurological disorder, medication use (or mis-use), or disordered substance use.

The duration and intensity of the jet lag symptoms are related to several factors: (1) the number of time zones crossed, (2) the direction of travel, (3) the ability to sleep while traveling, (4) the availability and intensity of local circadian time cues on arrival, and (5) individual differences in tolerance to circadian misalignment.[14]

Although jet lag is typically benign and self-limited, feeling alert, productive, and cheerful are important goals for travelers; thus, safe and effective treatment is often sought. Sometimes serious, unfortunate, and even dangerous consequences can result from jet lag, including untoward business, diplomatic, or military decisions. Athletes, who compete in international sports requiring optimal performance, have understandable concerns with the effects of jet lag. Some folks have pointed to the iconic Rose Bowl football game outcomes over the past 100 years as an indication of the disadvantage of jet lag in athletic competitions. The Rose Bowl is played each year in Pasadena, California, which means that the teams in the Pacific Coast Conference (Pac-12) are less likely than their competitors (non–West Coast teams) to be affected by significant jet lag at game-time. Accordingly, West Coast teams win the Rose Bowl more often (52 wins and 3 ties) than non–West Coast teams. Even racehorses travel across the United States and between the United States and Europe via specially equipped jets, and their performance is subject to jet lag![15] Among professional travelers such as flight personnel, diplomats, and international business executives, JLD may be recurrent or even chronic. In any case, jet lag provokes mood changes in some people and these changes can be profound.

The onset of mood disorders can often be triggered by disrupted sleep, and the role of jet lag in sleep disruption has been known for decades.[16] Jet lag seems to exacerbate psychotic conditions by disrupting the circadian system, and, among vulnerable individuals, it can even precipitate new psychosis.[17,18] Jet lag is associated with the greatest risk of provoking the symptoms of mood disorders when multiple time zones are crossed during the same trip. The severity and extent of the symptoms associated with jet lag also depend on the direction of travel. For the majority of people, it is much easier to adjust to a longer day (stay up later) than to adjust to a shorter day (earlier time of sleep onset); thus, westward travel generally provokes less pronounced jet lag than eastward travel.[19] Typically, the rate of adaptation to a new time zone is about half a day per hour of westward time shift versus about one day per hour of the time difference eastward. Thus, a trip to Paris from New York requires a 5-hour phase-advance shift. Typically, such a phase-shift requires approximately 5 days to readjust to local time—just in time to return back to New York! The good news is that only 2.5 or 3 days will be required to readjust to local time when making the same 5-hour phase-delay.

Among individuals with prior mood disorders, circadian rhythms may already be disrupted prior to jet travel; research has linked genetic polymorphisms (gene variants) or dysregulations of clock genes to bipolar disorder and schizophrenia.[19,20] Schizophrenia is a debilitating disorder of brain and behavior that affects thinking, feelings, and actions. People with schizophrenia find it difficult to distinguish fantasy from reality, and struggle with decision-making and managing emotions. Cognitive functions (i.e., thinking) may be disorganized, and perceptions are often unrelated to reality. For example, people with schizophrenia may hear voices and believe that others are reading their thoughts or controlling their thoughts. Symptoms generally begin in an individual's twenties and continue intermittently throughout life; these symptoms are generally terrifying to the people who are experiencing them, and common reactions include agitation, lack of responsiveness, or even total withdrawal. Although schizophrenia is a chronic disorder with no cure, the symptoms can be managed with medication and psychological treatments. However, many people find the side effects of the drugs too negative and stop taking the drugs after a while, allowing the symptoms to re-emerge.

One cause (or effect) of the relationship between sleep problems and schizophrenia is the quasi-nocturnal lifestyle adopted by many individuals with schizophrenia, which contributes to a limited exposure to appropriately timed light. There is a strong relationship between schizophrenia and sleep. Disrupted sleep and circadian rhythms have been reported for up to 80% of people with schizophrenia.[21] Disturbed sleep in people with schizophrenia reflects later sleep onset and reduced total sleep and slow wave sleep duration; onset of REM sleep is also delayed in this disorder.[19] Disrupted circadian rhythms are presumed because people with schizophrenia display atypical daily activity patterns including unstable and fragmented rest-activity rhythms.[10] This disorder also limits exposure to social zeitgebers, which are interpersonal events that regulate circadian rhythms and synchronize them with the environment.[22] For instance, in one study patients admitted to hospital for psychiatric emergencies in Hawaii were more likely to display depression or mania if they had recently traveled across multiple time zones compared with nontravelers.[23]

Social Jet Lag

The strong social temporal influences on human sleep become apparent by comparing the sleep-wake patterns on work/school and free days.[7] Most folks living in the United States or Europe display obvious differences in both sleep duration and timing.

When expressed as a weekly average, sleep duration dramatically reduces by nearly 2 hours between the ages of 10 and 17 years![9] Remarkably, the sharp reduction in sleep duration mainly reflects loss of sleep on school nights (>2 h), whereas sleep duration on free days becomes only gradually shorter from the ages of 10 to 65 years.[10] The profound mismatch between the "enforced" sleep duration on school/workdays and weekend/free-day sleep duration establishes the significant effect of social influences on daily sleep behavior. Furthermore, the mismatch disappears completely at around the average age of US and European retirement (i.e., 65 years of age).[10]

In one study of Korean workers, it was reported that ~10% of the population displayed weekly social jet lag >2 h. As the amount of social jet lag increased, the rate of depression also increased.[24] Similarly, a study of Japanese day-shift workers revealed that individuals who had >2 hours of social jet lag per week displayed a significantly higher rate of depressive symptoms.[25] The relationship between social jet lag and depressive symptoms appeared to be straightforward in this study. Similarly, a study of people in the Netherlands and two studies of adolescents in Brazil also concluded that social jet lag was associated with depression.[26-28] Taken together, these results suggest that disruption of circadian rhythms via social jet lag seems associated with depression across many cultures.

Exposure to Dim Lighting During the Day

We need bright light during the day. Increased daytime light exposure is now being used to successfully treat a range of depressive disorders. For example, in people with mild to severe depression, light treatment evokes a rapid remission rate ranging between 40% and 70%.[29,30] Among elderly people with dementia, sleep consolidation and behavioral disorders were improved by exposure to morning bright light therapy.[31] In another study, women working in offices with a window were exposed to higher light levels during the day and reported better sleep and lower depressive symptoms than women working in similar jobs, but without office windows.[32]

Daily bright light treatment is probably best known for treatment of winter depression, or seasonal affective disorder (SAD). SAD is characterized by depressed mood, lethargy, loss of libido, hypersomnia, anxiety, carbohydrate cravings, excessive weight gain, and the inability to focus attention or concentrate; these symptoms occur during the late autumn or winter.[32] In the Northern Hemisphere, symptoms usually begin between October and December and go into remission during March. These symptoms were previously considered simply as the "holiday blues"; however, this is not the case, because individuals with SAD in the Southern Hemisphere display symptoms six months out of phase with Northern Hemisphere residents (i.e., between April and June), which does not correspond to major holidays.[31] With the onset of the long days of summer, SAD patients regain their energy and become active and elated, often to the point of hypomania or mania. Three features atypical of depression—hyperphagia, carbohydrate cravings, and hypersomnia—set SAD apart from nonseasonal depression. SAD is often diagnosed as "bipolar II" depression or "atypical bipolar disorder," particularly if hypomania or mania is present.[1] Because of the relatively short duration of daylight during the fall and winter at higher latitudes, many people go to work or school in the dark, stay indoors under relatively dim lighting, then return home in the dark. Under such conditions, circadian rhythms can easily start to free-run (i.e., unlink from the solar day) and various internal processes dissociate from optimal.

The rates of SAD range from 1% to 10% of the population; higher prevalence rates are generally reported at high latitudes and lower prevalence is reported at low latitudes.[33] There is a sex difference in SAD; the sex ratio of SAD prevalence reported in one epidemiological study was 3.5 women to 1 man; more than 80% of respondents to newspaper advertisements recruiting experimental subjects with SAD were women.[33]

Improperly entrained circadian rhythms may be involved in SAD cases.[34] In other cases of depressed mood, it has been hypothesized that changing the onset of sleep time and social factors resets biological clocks, resulting in amelioration of the depression. Indeed, in one of the earliest studies of SAD, one depressed patient was phase-advanced in her sleep-wake cycle by 6 hours over several weeks; her depression was temporarily ameliorated by this treatment.[34]

The current standard of care for SAD is treatment with bright lights instead of sleep therapy. When patients are exposed to bright light, usually for a few hours during the morning, signs of remission of the SAD symptoms often become apparent within a few days.[35] Phototherapy, as sleep-wake therapy, may work by phase-advancing biological rhythms. Bright light has been suggested to possess two antidepressant effects: (1) light treatment in the morning may ameliorate depression by realignment of inappropriately entrained circadian rhythms, and (2) light may also serve as a general "energizer . . . of mood in a way that may be attributable wholly or in part to a placebo effect."[36] Light treatment in the evening has no mood benefits. Phototherapy appears to shift circadian rhythms by altering the timing of the nightly secretion of melatonin.[37]

In mammals, including humans, the nocturnal synthesis and secretion of melatonin from the pineal gland can be rapidly inhibited by exposure to brief periods of light at night.[36] Thus, light displays two vital actions in humans, as it does in other mammals: (1) light can entrain, or synchronize, the daily melatonin rhythm, and (2) light can acutely suppress daily melatonin secretion. Presumably, either or both effects of light may be involved in the therapeutic effects of phototherapy in the treatment of SAD.[37] The onset of melatonin secretion, which occurs about 14 hours after awakening, serves as a marker of circadian phase for synchronization of other circadian rhythms, and it may also influence circadian phase. For instance, men and women residing in dim light and given melatonin at specific times of the day shifted the onset time of their endogenous melatonin rhythms. Melatonin administered on four consecutive days during the late afternoon or early evening tended to cause the onset of the internal melatonin secretion to start earlier, whereas melatonin given in the morning tended to delay the onset of melatonin secretion.[37]

The results of additional studies suggest the existence of an approximately 12-hour period during which exogenous (external) melatonin treatment advances the internal biological rhythms, and a 12-hour period during which exogenous melatonin delays the endogenous cycle. This information allows for the therapeutic use of melatonin in resetting the phase of circadian rhythms. For example, the advance zone is usually between 6 and 18 hours after awakening, whereas the delay zone usually begins about 18 hours after awakening and continues during sleep to about 6 hours after awakening. Thus, light and melatonin can be given to treat ailments such as jet lag, problems associated with shift work, advanced and delayed sleep phase disorders, and free-running rhythms among blind people.

It has been estimated that about half the 200,000 totally blind people in the United States have free-running circadian rhythms.[38] This results in people whose circadian rhythms are not entrained to the 24-hour day and frequently results in insomnia during the night, daytime sleepiness, and other adverse effects.[39] This condition has been termed "non-24-hour sleep-wake disorder." In one study, several completely blind individuals displayed free-running periods ranging between 24.2 and 24.9 hours. After treatment with 10 mg/day of melatonin at bedtime, six out of seven individuals displayed entrained circadian rhythms.[40] A melatonin-like drug called Hetlioz (tasimelteon) has been marketed for treating non-24-hour sleep-wake disorder, but its effectiveness compared with over-the-counter melatonin supplements (available in the United States) remains unreported.

Humans require high-intensity illumination to suppress nighttime melatonin secretion.[39] Daytime illumination levels outdoors at temperate latitudes range between 25,000 and 100,000 lux,[41] whereas levels of artificial illumination indoors typically vary from 200 to 500 lux. Physiologically, however, humans respond quite differently to the high levels of illumination provided by exposure to sunlight. Exposure to at least 1,500 lux is necessary for the inhibition of human melatonin secretion.[42] This requirement may explain why typical indoor levels of artificial illumination are insufficient to relieve the symptoms of SAD; much brighter light must be used for effective treatment. Thus, people who develop SAD may possess defects in the light transduction pathways from the eyes to the primary circadian clock in the hypothalamus.[42] One study suggests that people suffering from SAD have a small genetic mutation in the melanopsin molecule; people diagnosed with SAD were 5.6 times more likely than people with no history of psychopathology to have a genetic mutation of melanopsin (rs2675703 [P10L]). Thus, genetic deficits in the nonvisual light input pathway from the eye to the central biological clock may represent an important mediator of SAD and point to additional effective treatments.[43,44] Indeed, people diagnosed with SAD showed an impaired postillumination pupil response compared with healthy people.[45] Individuals with variations in the melanopsin gene (OPN4) who may suffer from SAD may also display differences in alertness, circadian entrainment, and melatonin secretion.

At the level of genes associated with psychiatric disorders, several studies have identified correlations between schizophrenia and various clock genes (reviewed in[46]). For example, some evidence exists pointing to an association between the clock genes, *Per3* and *Tim*, and schizophrenia and schizoaffective disorder; *Cry1* has been suggested as a candidate gene for schizophrenia because it is located near a so-called hot spot for schizophrenia on a specific chromosome (12q24).[46,47]

Exposure to Light at Night

Elderly individuals institutionalized in hospitals or nursing facilities are particularly vulnerable to exposure to artificial light at night because 24-hour nursing activities and safety concerns require constant lighting. Circadian disruption is already a factor in the natural aging process, as nightly melatonin rhythms decrease in amplitude and sleep becomes more fragmented.[48] In a cohort of 857 elderly individuals, exposure to light at night was associated with insomnia and poor sleep quality,[49] which could be factors contributing to depressed mood. In another cohort of 516 elderly individuals, both intensity and duration of exposure to light at night were associated with depressive symptoms. Depressed individuals averaged more than 5 lux of light at night and more than 30 min of nightly exposure, compared with only 0.8 lux in the nondepressed group. Interestingly, depressive symptoms did not correlate with melatonin concentrations.[50] If it is feasible, then timing light exposure to a more appropriate circadian phase could be a strategy to reduce depression among the aged population. A number of studies on night-shift workers have suggested that exposure to light at night provokes depression (see next section). However, several other factors are affected by night-shift work that may also contribute to depressive symptoms.

Of course, the results from human studies are typically correlational and do not establish a causal relationship between disruption of circadian rhythms and mood disorders. As noted, many of the negative consequences of night-shift work have been attributed to disturbed sleep. There have been dozens of experimental studies in nocturnal rodents over the past two decades showing that the light at night disrupts circadian rhythms and changes affective (mood) responses. One benefit of examining nocturnal rodents includes separating the effects of light at night on circadian organization from sleep, as nocturnal rodents sleep during the day when it is typically light. Other benefits include direct experimental studies to test hypotheses regarding the effects of light at night on physiology and behavior. For example, dim light at night evokes depressive-like responses in rats, mice, and hamsters.[51-54]

You are probably asking yourself, how do you know that a rodent is depressed? First, when talking about nonverbal animal models of mood disorders, scientists refer to depressive-like responses, not clinical depression. Second, a number of symptoms associated with human depression can be studied. For example, rodents in a water-filled beaker will swim freely, but at some point, after trying to escape, start to float during these 5-minute tests. Increased immobility in this swim test is typically interpreted as behavioral despair, a symptom of human depression. Importantly, if this behavior is a valid representation of human depression, then treatment with an antidepressant drug ought to ameliorate the symptoms, and such treatment successfully reduces the depressive-like responses. Recall that in human depression the reduction of pleasurable activities is referred to as anhedonia. Because we cannot know what activities are more "pleasurable" in nonverbal animals, anhedonia is defined as individuals displaying fewer behaviors that were previously rewarding. For instance, anhedonia is often measured by examining the amount of sweetened water consumed as compared to plain water in a two-bottle choice test in mice. Again, treatment with an antidepressant drug ought to ameliorate the anhedonic symptoms, and many studies have reported that such drug treatments do reduce the anhedonic symptoms.

After only four weeks of exposure to light at night, hamsters increase depressive-like symptoms. That is, they display increased immobility in the swim test (despair) and reduced sucrose preference (anhedonia). Remarkably, within only two weeks of reinstatement of dark nights instead of light at night, the depressive-like behaviors disappear! If these results pertain to humans, then treatment of depression provoked by light at night may simply require elimination of the nightly light exposure. Also, two weeks of chronic treatment with a selective serotonin reuptake inhibitor (SSRI), citalopram, also ameliorates the depressive-like symptoms. One wonders how many people are taking Prozac, Zoloft, and other types of SSRIs to alleviate their depression induced by light-at-night exposure.

Further validation of these animal models involves the underlying mechanisms of depression. The hippocampus is one of the brain structures implicated in MDD. The name hippocampus is derived from the Greek *hippokampus* (*hippos*, meaning "horse," and *kampos*, meaning "sea monster") because the shape of the structure resembles a sea horse. Depressed individuals often display hippocampal shrinkage (atrophy) in brain scans, as well as dysregulated hippocampal functioning, including memory and stress responses. Similarly, the hippocampus is disproportionately vulnerable to neuroinflammation compared to other brain structures. Rodents exposed to artificial light at night display reduced hippocampal size and function after only four weeks of exposure. Changes in the structures of neurons (the brain cells that convey information throughout the brain) were observed after light at night exposure that may contribute to the reduction of hippocampal size and function. Exposure to light at night also ramped up neuroinflammation in the brain compared to the brains of animals housed under dark nights. Hippocampal levels of TNF gene products (one of the cytokines associated with inflammation) were increased after only four weeks of light at night exposure in concert with the increased levels of depressive-like behaviors. Treatment with a drug that inhibited TNF, however, blocked increased depressive-like behaviors after exposure to light at night.

One alternative treatment for depression provoked by light at night might be to manipulate the wavelength of light exposure. Recall that the retinal cells that transduce light information to the circadian clock are minimally responsive to long wavelength (red) light. Replacing standard bulbs with orange/red lights where possible, or using glasses that only pass red wavelengths, may be an effective approach to increasing visibility at night while preventing the disruptive effects of light at night. Preliminary studies in rodents that block out exposure to short wavelength (blue) light suggest that an environmental intervention could reap benefits in physiological and behavioral responses to light at night.

Rodent studies also revealed relationships between the circadian system and anxiety-like disorders. Again, animal studies of human mood disorders focus on specific components of the disorders. For instance, anxiety disorder is often assessed by the so-called open field test. This is a large box or circle into which the animals are placed. Spending most of the time in the device staying near the walls is an indicator of anxiety-like responses. If the animal ventures out into the center, the "open field," then it is considered to be displaying less anxiety-like behavior. Treatment of a mouse with valium, a commonly prescribed anxiolytic drug, is associated with increased time spent in the open field.

Targeted disruption of genes associated with the molecular clock influences anxiety-like behaviors. For instance, mice with a genetic mutation in the "Clock gene" display reduced anxiety-like responses and are less fearful of aversive stimuli than wildtype mice.[55] Mice lacking both versions of the Period gene display elevated anxiety-like behavior, whereas mice that lack either version of the Period gene do not display altered anxiety-like responses.[56] Inhibition of Period gene expression in the nucleus accumbens (NAc) of wildtype mice also produces anxiety-like behavior, suggesting a causal role for these core clock components. The NAc is a brain region that regulates anxiety in both humans and nonhuman animals.

Night-Shift Work

There are only a few studies that report that night-shift work is associated with MDD compared to their day shift counterparts. For example, the prevalence of MDD among night-shift workers was reported to be significantly increased relative to day-shift workers in a population of ~4,000 South Koreans.[57] Among a Brazilian cohort of ~36,000 workers, night-shift work was significantly associated with MDD, but only for women.[58] If all types of depression are combined, however, then a clear association emerges between night-shift work and depression.[59-62] A meta-analysis of 11 studies concluded that night-shift workers are ~40% more likely to develop depression relative to day-shift workers doing the same job and having the same socioeconomic status.[63]

Studies have also reported that individuals working night shifts display increased levels of anxiety compared to those with regular sleep-wake cycles.[64,65] Although several studies have suggested that night-shift work evokes anxiety, more recent analyses suggest that the mood changes may reflect disturbed sleep, rather than disturbed circadian rhythms per se. For instance, in one longitudinal study, day-shift workers without prior sleep disturbances who transitioned to rotating shift work schedules reported elevated anxiety along with disordered sleep.[65] Similarly, nurses with shift work disorder earned relatively high anxiety scores on the Hospital Anxiety and Depression Scale.[66]

As mentioned, what likely makes night-shift work so detrimental to health is that night-shift individuals experience many factors working against their circadian rhythms. Night-shift workers experience increased exposure to light, elevated caloric intake, and reduced sleep during their "night," which usually is attempted in a relatively bright bedroom illuminated by the sun. Night-shift workers who minimize the effects of artificial light at night by using blue light-blocking glasses potentially decrease their risk of depression.

To mimic night shifts in mouse models of disrupted circadian rhythms, exposure to light at night is often used. In many cases, adjustments to the environmental lighting cycles include models of jet lag (e.g., 1–8-hour phase shifts—i.e., turning on or off the lights at different times), light-dark inversions (e.g., a 12-hour phase shift for animals maintained on 12 hours of light and 12 hours of dark day [LD 12:12]), or nonentraining 24-hour days (e.g., 20- or 28-hour days). These sorts of mouse studies suggest that dysregulated circadian rhythms evoke depressive-like responses.

Most successful treatments of depression in humans— bright light therapy, wake therapy, social rhythm therapy, and antidepressants (SSRIs, SNRIS, and agomelatine)—directly affect circadian rhythms.[67] Morning bright light therapy phase advances the nightly melatonin rhythm and the degree of phase advancement is correlated with the amelioration of depressive symptoms.[68] Additionally, administration of antidepressants (SSRIs, SNIRs, and agomelatine [melatonin receptor mimics]) result in a phase advancement of circadian rhythms (i.e., melatonin secretion, body temperature, activity, and cortisol rhythms).[69,70] Another treatment approach called "wake therapy" directly alters sleep-wake rhythms by increasing the amount of slow wave sleep (SWS), reducing the latency to REM sleep (rapid eye movement is when dreaming occurs), and reducing the duration of REM sleep.[71] Thus, successful treatment of MDD with chronotherapies, as well as time-of-day variation in symptoms, provides evidence that circadian disruption may underlie this disorder.

Considered together, both clinical and preclinical (nonhuman animal) studies provide strong evidence that disrupted circadian rhythms may initiate, maintain, or worsen mood disorders. The interaction of sleep disorders with disrupted circadian rhythms is critical and will be explored in Chapter 4.

References

1. American Psychiatric Association. 2013. *Diagnostic and Statistical Manual of Mental Disorders*. 5th ed. American Psychiatric Association, Arlington, VA.

2. World Health Organization. 2017. Depression and other common mental disorders: global health estimates (No. WHO/MSD/MER/2017.2). World Health Organization.

3. Hidaka BH. 2012. Depression as a disease of modernity: explanations for increasing prevalence. *Journal of Affective Disorders*, 140: 205–214.

4. Egeland JA, Horstetter AM. 1983. Amish study. I: Affective disorders among the Amish, 1976–1980. *American Journal of Psychiatry*, 140: 56–61.

5. Emens J, Lewy A, Kinzie JM, Arntz D, Rough J. 2009. Circadian misalignment in major depressive disorder. *Psychiatry Research*, 168: 259–261.

6. Li JZ, et al. 2013. Circadian patterns of gene expression in the human brain and disruption in major depressive disorder. *Proceedings of the National Academies of Science*, 110: 9950–9955.

7. Germain A, Kupfer DJ. 2008. Circadian rhythm disturbances in depression. *Human Psychopharmacology*, 23: 571–585.

8. Gbyl K, Østergaard Madsen H, Dunker Svendsen S, Petersen PM, Hageman I, Volf C, Martiny K. 2016. Depressed patients hospitalized in southeast-facing rooms are discharged earlier than patients in northwest facing rooms. *Neuropsychobiology*, 74: 193–201.

9. Benedetti F, Colombo C, Barbini B, Campori E, Smeraldi E. 2001. Morning sunlight reduces length of hospitalization in bipolar depression. *Journal of Affective Disorders*, 62:221–223.

10. Foster RG, Peirson SN, Wulff K, Winnebeck E, Vetter C, Roenneberg T. 2013. Sleep and circadian rhythm disruption in social jetlag and mental illness. *Progress in Molecular and Biological Translational Sciences*, 119:325–346.

11. Angst J, Vollrath M. 1991. The natural history of anxiety disorders. *Acta Psychiatric Scandinavia*, 84: 446–452.

12. Walker WH, Walton JC, DeVries AC, et al. 2020. Circadian rhythm disruption and mental health. *Translational Psychiatry*, 10: 28.

13. American Academy of Sleep Medicine. 2005. *The International Classification of Sleep Disorders: Diagnostic and Coding Manual*. 2nd ed. American Academy of Sleep Medicine. Westchester, IL.

14. Sack RL. 2009. The pathophysiology of jet lag. *Travel Medicine and Infectious Disease*, 7: 102–110.

15. Elliott JA, Nelson RJ. 2010. It's about time: the coupling of biological clocks and veterinary medicine. *Veterinarian Journal*, 185: 98–99.

16. Flinn DE, Gaarder KR, Smith DC. 1959. Acute psychotic reactions during travel. *US Armed Forces Medical Journal*, 10: 524–531.

17. Perrier E, Manen O. 2011. [Jet lag disorders]. *Reviews Medicine Interne*, 32: S233–S235.

18. Sack RL. 2010. Jet lag. *New England Journal of Medicine*, 362: 440–447.

19. Mansour HA, Wood J, Logue T, et al. 2006. Association study of eight circadian genes with bipolar I disorder, schizoaffective disorder and schizophrenia. *Genes Brain and Behavior*, 5: 150–157.

20. Takao T, Tachikawa H, Kawanishi Y, Mizukami K, Asada T. 2007. CLOCK gene T3111C polymorphism is associated with Japanese schizophrenics: a preliminary study. *European Neuropsychopharmacology*, 17: 273–276.

21. Cohrs S. 2008. Sleep disturbances in patients with schizophrenia: impact and effect of antipsychotics. *CNS Drugs*, 22: 939–962.

22. Mistlberger RE, Skene DJ. 2004. Social influences on mammalian circadian rhythms: animal and human studies. *Biological Reviews of the Cambridge Philosophical Society*, 79: 533–556.

23. Young DM. 1995. Psychiatric morbidity in travelers to Honolulu, Hawaii. *Comprehensive Psychiatry*, 36: 224–228.

24. Min J, Jang TW, Lee HE, Cho SS, Kang MY. 2023. Social jetlag and risk of depression: results from the Korea National Health and Nutrition Examination Survey. *Journal of Affective Disorders*, 323: 562–569.

25. Islam Z, Hu H, Akter S, Kuwahara K, Kochi T, Eguchi M, Kurotani K, Nanri A, Kabe I, Mizoue T. 2020. Social jetlag is associated with an increased likelihood of having depressive symptoms among the Japanese working population: the Furukawa Nutrition and Health Study. *Sleep*, 43: 204.

26. de Souza CM, et al. 2014. Midpoint of sleep on school days is associated with depression among adolescents. *Chronobiology International*, 31: 199–205.

27. Levandovski R, et al. 2011. Depression scores associate with chronotype and social jetlag in a rural population. *Chronobiology International*, 28: 771–778.

28. Knapen SE, et al. 2018. Social jetlag and depression status: results obtained from the Netherlands Study of Depression and Anxiety. *Chronobiology International*, 35: 1–7.

29. Golden RN, Gaynes BN, Ekstrom RD, et al. 2005. The efficacy of light therapy in the treatment of mood disorders: a review and meta-analysis of the evidence. *American Journal of Psychiatry*, 162: 656–662.

30. Martiny K, Lunde M, Unden M, Dam H, Bech P. 2005. Adjunctive bright light in non-seasonal major depression: results from clinician-rated depression scales. *Acta Psychiatry Scandinavia*, 112: 117–124.

31. Mishima K, Okawa M, Hishikawa Y, Hozumi S, Hori H, Takahashi K. 1994. Morning bright light therapy for sleep and behavior disorders in elderly patients with dementia. *Acta Psychiatry Scandinavia*, 89: 1–7.

32. Harb F, Hidalgo MP, Martau B. 2015. Lack of exposure to natural light in the workspace is associated with physiological, sleep and depressive symptoms. *Chronobiology International*, 32: 368–375.

33. Rosenthal NE. 1993. *Winter Blues: Seasonal Affective Disorder*. Guilford Press, New York.

34. Rosenthal NE, Sack DA, Skwerer RG, Jacobsen FM, Wehr TA. 1988. Phototherapy for seasonal affective disorder. *Journal of Biological Rhythms*, 3: 101–120.

35. Terman M. 1988. On the question of mechanism in phototherapy for seasonal affective disorder: considerations for clinical efficacy and epidemiology. *Journal of Biological Rhythms*, 3: 155–172.

36. Lewy AJ, Sack RL, Singer CM, White DM, Hoban TM. 1988. Winter depression and the phase-shift hypothesis for bright light's therapeutic effects: history, theory, and experimental evidence. *Journal of Biological Rhythms*, 3: 121–134.

37. Lewy AJ, Emens JS, Songer J, Rough J. 2009. The neurohormone melatonin as a marker, medicament and mediator. In: *Hormones, Brain and Behavior*, DW Pfaff et al. (Eds.), pp. 2505–2526. Academic Press, San Diego.

38. Lewy AJ, Bauer VK, Hasler BP, Kendall AR, Pires ML, Sack RL. 2001. Capturing the circadian rhythms of freerunning blind people with 0.5 mg melatonin. *Brain Research*, 918: 96–100.

39. Lewy AJ, Emens J, Sack RL, Hasler BP, Bernert RA. 2003. Zeitgeber hierarchy in humans: resetting the circadian phase positions of blind people using melatonin. *Chronobiology International*, 20: 837–852.

40. Sack RL, Lewy AJ. 2001. Circadian rhythm sleep disorders: lessons from the blind. *Sleep Medicine Reviews*, 5: 189–206.

41. Brainard GC, Richardson BA, King TS, Matthews SA, Reiter RJ. 1983. The suppression of pineal melatonin content and N-acetyltransferase activity by different light irradiances in the Syrian hamster: a dose-response relationship. *Endocrinology*, 113: 293–296.

42. Lewy AJ, Wehr TA, Goodwin FK, Newsome DA, Markey SP. 1980. Light suppresses melatonin secretion in humans. *Science*, 210: 1267–1269.

43. Roecklein KA, Wong PM. 2020. Seasonal affective disorder. In: *Encyclopedia of Behavioral Medicine*, MD Gellman (Ed.), pp. 1722–1724. Springer, Cham. https://doi.org/10.1007/978-3-030-39903-0_836.

44. Roecklein KA, Wong PM, Miller MA, Donofry SD, Kamarck ML, Brainard GC. 2013. Melanopsin, photosensitive ganglion cells, and seasonal affective disorder. *Neuroscience and Biobehavioral Reviews*, 37: 229–239.

45. Roecklein K, Wong P, Ernecoff N, Miller M, Donofry S, Kamarck M, Wood-Vasey WM, Franzen P. 2013. The post illumination pupil response is reduced in seasonal affective disorder. *Psychiatry Research*, 210: 150–158.

46. Lamont EW, Coutu DL, Cermakian N, Boivin DB. 2010. Circadian rhythms and clock genes in psychotic disorders. *Israeli Journal of Psychiatry and Relationship Sciences*, 47: 27–35.

47. Logan RW, McClung CA. 2019. Rhythms of life: circadian disruption and brain disorders across the lifespan. *Nature Reviews Neuroscience*, 20: 49–65.

48. Pandi-Perumal SR, Zisapel N, Srinivasan V, Cardinali DP. 2005. Melatonin and sleep in aging population. *Experimental Gerontology*, 40: 911–925.

49. Obayashi K, Saeki K, Kurumatani N. 2014. Association between light exposure at night and insomnia in the general elderly population: the HEIJO-KYO cohort. *Chronobiology International*, 31: 976–982.

50. Obayashi K, Saeki K, Iwamoto J, Ikada Y, Kurumatani N. 2013. Exposure to light at night and risk of depression in the elderly. *Journal of Affective Disorders*, 51: 331–336.

51. Bedrosian TA, Weil ZM, Nelson RJ. 2012. Chronic citalopram treatment ameliorates depressive behavior associated with light at night. *Behavioral Neuroscience*, 126: 654–658.

52. Bedrosian TA, Fonken LK, Walton JC, Haim A, Nelson RJ. 2010. Dim light at night provokes depression-like behaviors and reduces CA1 dendritic spine density in female hamsters. *Psychoneuroendocrinology*, 36: 1062–1069.

53. Bedrosian TA, Weil ZM, Nelson RJ. 2013. Chronic dim light at night provokes reversible depression-like phenotype: possible role for TNF. *Molecular Psychiatry*, 18: 930–936.

54. Bedrosian TA, Vaughn CA, Galan A, Daye G, Weil ZM, Nelson RJ. 2013. Nocturnal light exposure impairs affective responses in a wavelength-dependent manner. *Journal of Neuroscience*, 33: 13081–13087.

55. Roybal K, et al. 2007. Mania-like behavior induced by disruption of CLOCK. *Proceedings of the National Academy of Sciences*, 104: 6406–6411.

56. Spencer S, et al. 2013. Circadian genes Period 1 and Period 2 in the nucleus accumbens regulate anxiety-related behavior. *European Journal of Neuroscience*, 37: 242–250.

57. Min J, Jang TW, Lee HE, Cho SS, Kang MY. 2023. Social jetlag and risk of depression: results from the Korea National Health and Nutrition Examination Survey. *Journal of Affective Disorders*, 323: 562–569.

58. Oenning NSX, Ziegelmann PK, De Goulart BNG, Niedhammer I. 2018. Occupational factors associated with major depressive disorder: a Brazilian population-based study. *Journal of Affective Disorders*, 240: 48–56.

59. Moon HJ, Lee SH, Lee HS, Lee KJ, Kim JJ. 2015. The association between shift work and depression in hotel workers. *Annals Occupational Environmental Medicine*, 27: 29.

60. Lee HY, Kim MS, Kim O, Lee IH, Kim HK. 2016. Association between shift work and severity of depressive symptoms among female nurses: the Korea Nurses'. Health Study. *Journal of Nursing Management*, 24: 192–200.

61. Booker LA, et al. 2019. Exploring the associations between shift work disorder, depression, anxiety and sick leave taken amongst nurses. *Journal of Sleep Research*, e12872.

62. Knapen S, et al. 2018. Social jetlag and depression status: results obtained from Netherlands Study of Depression and Anxiety. *Chronobiology International*, 35: 1–7.

63. Lee A, et al. 2017. Night shift work and risk of depression: meta-analysis of observational studies. *Journal of Korean Medical Sciences*, 32: 1091–1096.

64. Gamble KL, Motsinger-Reif AA, Hida A, Borsetti HM, Servick SV, et al. 2013. Shift work in nurses: contribution of phenotypes and genotypes to adaptation. *PLoS One*, 8: e70423.

65. Kalmbach DA, Pillai V, Cheng P, Arnedt JT, Drake CL. 2018. Shift work disorder, depression, and anxiety in the transition to rotating shifts: the role of sleep reactivity. *Sleep Medicine*, 52: 162–166.

66. Flo E, et al. 2012. Shift work disorder in nurses: assessment, prevalence and related health problems. *PLoS One*, 7: e33981.

67. Germain A, Kupfer DJ. 2008. Circadian rhythm disturbances in depression. *Human Psychopharmacology*, 23: 571–585.

68. Terman JS, Terman M, Lo ES, Cooper TB. 2001. Circadian time of morning light administration and therapeutic response in winter depression. *Archives of General Psychiatry*, 58: 69–75.

69. Leproult R, Van Onderbergen A, L'Hermite-Balériaux M, Van Cauter E, Copinschi G. 2005. Phase-shifts of 24-h rhythms of hormonal release and body temperature following early evening administration of the melatonin agonist agomelatine in healthy older men. *Clinical Endocrinology*, 63: 298–304.

70. Robillard R, et al. 2018. Parallel changes in mood and melatonin rhythm following an adjunctive multimodal chronobiological intervention with agomelatine in people with depression; a proof of concept open label study. *Frontiers in Psychiatry*, 9: 624.

71. Berger MV, Van Calker D, Riemann D. 2003. Sleep and manipulations of the sleep–wake rhythm in depression. *Acta Psychiatry Scandinavia*, 108: 83–91.

4

LIGHT AND SLEEP

Of the many circadian rhythms displayed by individuals, the sleep-wake cycle is probably the most salient to us. Many studies have reported a relationship between disrupted circadian rhythms and sleep.[1] Importantly, the extent to which disrupted sleep is the primary outcome (or cause) of downstream effects of disrupted circadian rhythms (e.g., metabolism, cardiovascular, or cancer) or whether disrupted circadian clock function is the primary outcome (or cause) of sleep disorders also remains somewhat unknown for humans.[2] Chapter 4 will review the relationship between disrupted circadian rhythms and sleep as currently understood.

Sleep is a ubiquitous behavior observed in virtually all animal species. However, the biological functions of sleep are not fully understood, and its evolutionary history is also a subject of ongoing research.[3] Sleep is a complex behavior that serves several physiological and psychological functions.[4] One of the primary functions of sleep is restoration and recovery of our bodies from the previous day(s) of activities. During sleep, we undergo several processes that help to repair and restore tissues, including the

Dark Matters. Randy J. Nelson, Oxford University Press. © Oxford University Press (2025).
DOI: 10.1093/9780197639979.003.0004

release of growth hormone and the synthesis of new proteins. Sleep also seems to be important for clearing the brain of cellular waste products through the enhanced circulation of cerebral spinal fluid[5]—failure of this function allows toxic proteins to accumulate, which in turn can affect neuronal health. Although the nature and functions of sleep remain an area of intense research, there is little doubt that sleep plays a critical role in memory consolidation and the integration of new information.[6]

The evolutionary history of sleep remains under study, but it is believed that sleep evolved in response to several selective pressures. One hypothesis of the role of sleep suggests that sleep evolved as a mechanism to conserve energy, although this biological advantage had to counter the costs of allowing individuals to be vulnerable to predation during periods of low activity. Another hypothesis proposes that sleep evolved to allow for neural development and repair, as sleep is especially important for the developing brain. We must assume that the function of sleep is critically important because many versions of sleep have evolved, virtually all animals engage in sleep, and chronic sleep deprivation can have fatal consequences. Whatever the selective pressures that led to the evolution of sleep, it has become apparent that sleep has become an essential behavior for almost all animal species.[7,8] There is broad diversity of sleep among different species. Individuals of some species sleep for only a few hours per day, whereas others sleep for up to 20 hours per day. Of course, individuals of some species sleep primarily during the night (nocturnal), whereas individuals of other species sleep mostly during the day (diurnal). In both diurnal and nocturnal animals, melatonin is exclusively secreted from the pineal gland during the dark of night.

There are two likely mechanisms that promote sleep: the sleep pressure hypothesis and the circadian rhythm hypothesis. The "sleep pressure hypothesis" suggests that the homeostatic drive for sleep is proportional to the duration of wakefulness. Adenosine, a neuromodulator, accumulates in the brain during wakefulness and is thought to contribute to sleep pressure. As adenosine concentrations increase, there is an increase in sleepiness, and when it reaches a certain threshold, it triggers the onset of sleep. During sleep, adenosine levels decrease, resetting the sleep pressure. Not surprisingly to coffee drinkers, caffeine inhibits accumulation of adenosine and thus promotes wakefulness. The "circadian hypothesis" of sleep suggests that the timing of sleep is regulated by circadian rhythms, and for most humans sleep occurs during the biological night. That is, a circadian-mediated signal promotes sleep onset. Circadian rhythms regulate the timing of various physiological functions, including the release of melatonin, a hormone that promotes sleepiness. Melatonin concentrations increase during the evening and peak during the biological night, promoting the onset of sleep in humans. Disrupted circadian rhythms can lead to a delay or advancement of the biological clock, resulting in a mismatch between the sleep-wake cycle and the internal circadian rhythms. As we'll see later in this chapter, disruptions in circadian rhythms and sleep can have significant consequences for health and well-being. The circadian and sleep pressure hypotheses provide a framework for understanding the regulation of sleep and the impact of disruptions in sleep patterns, and likely interact to promote healthy sleep patterns. Maintaining a regular sleep-wake cycle and synchronizing it with your natural circadian rhythms can help promote healthy sleep and prevent adverse health effects.[9,10]

Human sleep is divided into two main categories: non-rapid eye movement (NREM) sleep and rapid eye movement (REM) sleep. NREM sleep is typically divided into three stages, with each stage characterized by progressively deeper levels of sleep. REM sleep occurs approximately 90 minutes after falling asleep and recurs in ~90 minute cycles. REM sleep is characterized by rapid eye movements and is often associated with dreaming.

Despite its importance, many if not most individuals do not prioritize sleep and may experience chronic sleep deprivation. Just like your grandmother probably told you—sleep is very important for good health. As noted, sleep is essential for many physiological processes, including the regulation of metabolism, hormone secretion, and immune function. Adequate sleep has been associated with a reduced risk of various health conditions, including obesity, cardiovascular disease, diabetes, and depression. Additionally, sleep is necessary and important for cognitive function (thinking), making memories (memory consolidation), and regulation of emotions.

We all need to prioritize and promote healthy sleep habits. Thus, it is essential to prioritize good sleep hygiene. Strategies such as maintaining a consistent sleep-wake schedule (avoid social jet lag), limiting exposure to bright light during the night (especially blue light), and avoiding stimulants such as caffeine and nicotine can all help to promote healthy sleep, and healthy circadian rhythms. Additionally, engaging in regular exercise and relaxation techniques, such as meditation or yoga, can help to reduce stress and promote relaxation and improve sleep, not to mention overall health and well-being. Again, the circadian hypothesis of sleep suggests that the timing of sleep is regulated by circadian rhythms, and sleep occurs during the biological night. If our circadian clock mediates sleep-promoting signals, then it seems reasonable to propose that disrupting our circadian clocks might negatively affect sleep. For the most part, this seems to be true and the evidence that disrupted circadian rhythms impair sleep is presented in the following sections.

Jet Lag

Although jet lag is a common problem faced by people who travel across multiple time zones, jet lag, per se, is not experienced by travelers on north-south flights within the same time zone. Thus, jet lag does not merely reflect the exhausting aspects of air travel.[11] Jet lag can evoke a number of negative effects, including sleep disturbances, fatigue, and impaired cognitive performance (Table 4.1). Jet lag can have a significant impact on sleep, both in terms of its quality and duration; indeed, disturbed sleep is the most commonly reported symptom associated with jet lag. When traveling across multiple time zones, circadian rhythms become desynchronized, leading to difficulty falling asleep, staying asleep, and waking up at the appropriate times. For example, our body temperatures are programmed by our biological clocks to decrease during the evening. This reduction in body temperature promotes sleep onset. However, if body temperature instead peaks in the evening because these rhythms remain entrained (synchronized) to the previous time zone,

Table 4.1 Symptoms of Jet Lag.[12]

- Impaired sleep
- Gastrointestinal effects including: Heartburn, Constipation, Diarrhea, Nausea, Aversions to specific foods
- Fatigue
- Reduced awareness
- Confusion
- Anxiety
- Irritability
- Impaired memory
- Dizziness
- Disorientation
- Headaches
- Perspiration
- Muscle pains
- Irregular menstrual cycles in women who travel frequently
- General malaise

---

then sleep onset can be delayed. The same misaligned rhythms in cardiac function, metabolism, or endocrine function can also interfere with sleep. Thus, the desynchronization of the "hands of the circadian clock" can result in sleep deprivation, which can lead to a variety of negative effects on both physical and mental health.

One study indicated that individuals who experienced jet lag reported poorer sleep quality, longer sleep onset latency, and decreased sleep duration compared to individuals who did not experience jet lag.[13] Additionally, individuals who experienced jet lag reported increased daytime sleepiness and impaired cognitive performance, such as decreased attention, reaction time, and memory performance.[14] These results apply to both daytime and nighttime travels.

For most people, it is generally easier to travel across multiple time zones in a westward direction compared to eastward travel. This is because it is typically easier for people to stay up later (phase delay) than go to bed earlier (phase advance). For instance, one study examined westbound versus eastbound travel on sleep, subjective jet lag symptoms, and athletic performance in university athletes.[15] College athletes were tested on several physical tests important for team sports performance. Previous research in both athletes and nonathletes had demonstrated that grip strength was affected by jet lag.[16,17] Typically, there is a performance rhythm with a morning low-point and a late afternoon peak in grip strength. However, after transmeridian travel, grip strength is reduced outside of this time-of-day window. The study that focused on athletic performance skills necessary for team sports tested male students at two times per day (0900h and 1700h) on four consecutive days for two weeks prior to travel to obtain baseline information. After a flight across eight time zones from Sydney, Australia to Qatar (i.e., westbound), the men were retested for four days. They then returned (i.e., eastbound) across the eight time zones from Qatar to Sydney. In addition to physical data, sleep was assessed via activity monitors worn on the wrists as well as self-report diaries throughout the experiment. In addition to poorer physical results among the eastbound travelers, the average times of sleep onset and awakening were significantly later. The mean time spent in bed and the total sleep duration were significantly reduced across the four days among the students traveling eastbound. Taken together, it is apparent that transmeridian travel across multiple time zones can impair team sport physical performance. Although transient changes in grip strength may not be consequential for those of us who are not college athletes, the study establishes that eastbound travel increases negative effects on sleep, fatigue, and overall malaise.

There are many strategies that can be employed to minimize the negative effects of jet lag on sleep. One such strategy is to adjust sleep schedules prior to travel in order to align with the destination time zone. This can be achieved by gradually shifting sleep and wake times in the days leading up to travel. Another strategy is to strategically use light exposure to help reset the circadian clock. This can be achieved by exposing oneself to bright light in the morning in the destination time zone and avoiding bright light in the evening. Melatonin supplements may also be effective in promoting sleep and reducing the effects of jet lag, although more research is needed to determine and optimize its efficacy.[18]

Social Jet Lag

Social jet lag can have a significant effect on sleep, in terms of both its quality and duration.[19,20] When individuals' sleep schedule is misaligned with their internal circadian rhythms, people report difficulty falling asleep, staying asleep, and waking up at the appropriate times. The sleep of people who were experiencing social jet lag, especially evening chronotypes, tends to be shortened and of poor quality during school/work days.[21,22] Specifically, social jet lag can cause a sleep debt or even evoke chronic sleep deprivation, poor sleep quality, and impaired cognitive performance.[19] For instance, one study concluded that individuals who experienced social jet lag had poorer sleep quality, longer sleep onset latency, and decreased sleep duration compared to individuals who did not experience social jet lag.[23] Additionally, individuals who experienced social jet lag were more likely to report symptoms of daytime sleepiness and impaired cognitive performance, such as decreased attention, reaction time, and memory compared to people who did not experience social jet lag.[24]

Impaired sleep caused by social jet lag has a number of serious consequences, especially for mental health.[25,26] For example, one study linked nonsuicidal self-injury among college students with social jet lag and disrupted sleep.[27] Other research has linked poor sleep quality and social jet lag with substance abuse among European and US college students,[28] and depressive symptoms among Chinese college students.[29] Markedly, social jet lag decreased and sleep duration increased among adolescents during the COVID-19 pandemic as lockdowns prevented leaving home at night to socialize.[30]

Social Jet Lag Has Serious Consequences for Academic and Job Success

In one study, nearly 800 adolescents provided information about their sleep habits, morningness-eveningness, cognitive abilities, and grade point averages.[31] The amount of time in bed during the weekends was not related to cognitive abilities; only the weekday time in bed was related to academic achievement. Social jetlag was negatively related to academic achievement and general cognitive abilities. When the researchers examined the data more closely, they discovered that social jetlag seemed more detrimental to girls' performance than boys' performance.[31] Additional studies of the effects of social jet lag on cognitive performance will be reviewed in Chapter 5.

As we will learn in more detail in Chapter 8, addressing social jet lag requires implementing strategies to optimize sleep patterns. Maintaining a consistent sleep schedule, even on weekends and during social events, can help regulate our circadian rhythms and mitigate the effects of social jet lag on sleep. Exposure to natural light in the morning and limiting exposure to artificial light after dark can help to regulate circadian rhythms. Additionally, avoiding stimulants, such as caffeine, or other drugs such as alcohol, close to bedtime can enhance sleep quality and facilitate the adjustment to a regular sleep schedule.

Exposure to Dim Lighting During the Day

Although it might seem counterintuitive that our dimly lit daytime environments are not conducive to sleep later at night, the lack of bright days influences our daily rhythms and sleep and typically not in an optimal manner. Natural sunlight, typically between 25,000 and 100,000 lux, has been reported to: (1) advance the timing of sleep to earlier hours, (2) extend the duration of sleep, and (3) improve sleep quality.[32] For example, the phase-advancing effects of daylight based on questionnaire data were reported in one study; the researchers reported that for each additional hour spent outdoors the timing of sleep onset advanced by ~30 minutes.[22] Bright light exposure during the day affects sleep duration. In this study, reduced daylight exposure and prolonged nights are associated with an extended biological night as determined by the duration of nightly melatonin secretion; recall that melatonin is secreted only at night and is typically related to sleep onset and possibly sleep duration.[32]

Exposure to bright daylight has also been reported to increase sleep duration by advancing the onset time of sleep.[33] In a series of experiments to test the role of bright light exposure (and lack of artificial light at night) on sleep, Kenneth Wright and colleagues studied sleep in people in the lab and also during camping trips in the Rocky Mountains in Colorado. People on the camping trips were exposed to roughly 4 to 10 times the light levels during the day compared to what is typical in constructed environments. Importantly, time of sleep onset each night among the campers was about 2.5 hours earlier than when they were exposed to artificial light at night. Because they awoke at approximately the same time in the morning in both situations, their average sleep duration while camping was about 2.3 hours longer when exposed to natural sunlight as compared to living in a modern electrical lighting environment. Essentially, the variation in the duration of melatonin secretion and sleep times was minimized among those camping trips when they were exposed to the potent bright light during the day that synchronized their circadian rhythms.

However, for those of us who work inside during the day, a number of studies indicated that we can improve our sleep quality by increasing light exposure during the day in our homes and offices. It is important to recognize that all white light sources are not the same: some, such as LED and fluorescent light, radiate much more energy than others in the blue portions of the spectrum. These types of light are referred to as blue enriched white light. For example, daytime exposure to bright blue enriched white light is associated with increased evening fatigue at bedtime.[34] Importantly, daytime exposure to blue enriched light also improves sleep quality,[35] decreases the amount of time necessary to fall asleep,[36] and increases the amount of restful slow-wave sleep during the night, which is associated with reduced sleepiness during the day.[37] (For me personally, a sunny day at the beach helps me fall asleep more quickly and increases how long I sleep that night.) Taken together, the research to date suggests that bright daytime light (especially blue-wavelength-enriched) is beneficial for sleep.

Exposure to Light at Night

We are exposed to significant light pollution in modern societies. For instance, light intensities on a typical urban street range between 5 and 15 lux during the night, and a typical living room is illuminated nightly between 100 and 300 lux; electronic tablets emit ~40 lux, depending on the size of the screen.[38] Remarkably, 36% of adults and 34% of children surveyed sleep nightly with a light-producing electronic device, such as a computer or television.[39] The pervasiveness and intensity of nighttime light exposure is unprecedented in our history. Several studies suggest that exposure to artificial light at night, either from outdoor or indoor sources, negatively affects both subjective and objective sleep quality as indicated by actigraphy (measurement of movement) or polysomnography (a study of sleep that uses electroencephalogram (EEG), eye movements, muscle tone, and oxygen levels to monitor sleep. Indeed, light at night can evoke reduced total sleep time, sleep efficiency, increased awakenings after sleep onset, reduced amount of slow wave sleep, and increased arousal index (awakening from sleep).[40]

Over the past decade or so, light from LED screens, which consists of blue-enriched light, has been suggested to interfere with sleep.[32] Sleep-associated processes such as melatonin secretion are also altered by exposure to LED screens used before bedtime.[41] For instance, reading an eReader for four hours prior to sleep delays the onset of sleep, reduces evening sleepiness, reduces melatonin secretion, and next-morning alertness, and provokes phase delays of circadian rhythms.[42]

Several studies have reported that smartphone ownership and use before bedtime may be associated with more self-reported sleeping problems. For example, teenagers who owned a smartphone slept less than their peers who did not have a smartphone.[43] The negative effects of smartphones are not limited to adolescents. There was a strong correlation in adults between hours on their phones and decreased sleep efficiency, longer sleep onset latency, and poor sleep quality.[44] Smartphone use has also been demonstrated to delay sleep onset and reduced sleep duration.[32] One factor from these devices is certainly the light emitted by their screens, but the extent to which other factors associated with these devices, such as general arousal and psychological or emotional effects, remains unspecified.

In response to mounting evidence that visual display units such as eReaders, tablets, and smartphones could negatively affect sleep, makers of most current smartphones and tablets now provide a so-called night shift feature. This feature changes the spectral tone of the display during the evening hours toward longer (more reddish) wavelengths. Using the "night shift" mode reduces melanopsin activation by 67% at full display brightness. Although this may seem to be a significant reduction in potential disruptive light signaling to the circadian system, you can dim your smartphone to its minimum level and reduce melanopsin activation to less than 1% of activation compared to maximum display brightness.[32] Recent studies that used so-called metameric displays that do not differ in their appearance, but rather differ only in the extent of melanopsin stimulation, indicate that the nonvisual properties of light can be modulated independently of visual appearance.[45,46] These results suggest that using this approach may provide a solution to disrupted circadian rhythms by electronic devices in the future.

Night-Shift Work

The majority of studies linking disrupted circadian rhythms and poor sleep outcomes come from night-shift work. Night-shift workers are more susceptible to developing sleep disorders such as insomnia, shift work sleep disorder (SWSD), and excessive daytime sleepiness.[47] SWSD, specifically, is characterized by difficulty adjusting to a shift work schedule and experiencing insomnia or excessive sleepiness when desired sleep is not possible. At least 75% of night-shift workers report disturbed sleep! Because circadian rhythms influence the timing and quality of sleep, disrupting these rhythms has profound effects on sleep.[48]

People working night shifts sleep less during the day than they might during the night. For example, sleep ends after only 4–6 hours for night-shift workers; individuals on night shift report being unable to return to sleep. Yes, daytime sleep is 1–4 hours shorter than nighttime sleep!

This state of affairs likely reflects the notion that the circadian clock is promoting wakefulness during the induced circadian misalignment. The extensive sleep loss is primarily stage 2 sleep (the dominant sleep stage) and REM sleep (dream sleep). Sleep stages 3 and 4 (so-called deep sleep) seem unaffected. Night-work shifts are the worst shift for sleep, evoking greater loss of total sleep time than evening and slow rotating shift schedules. During slow rotating night-shift schedules, employees move between day and night shifts after longer periods such as after several weeks or months. Although some have argued that permanent night work may have benefits in terms of a more "long-term" circadian rhythm adjustment to night-shift work, there is little to no support for this argument.[49]

Strict control over exposure to light and dark can facilitate complete or partial adaptation to permanent night work schedules. That means using sleep masks and blackout curtains in the bedroom during the day and relatively bright light during the night shift. Application of circadian principles to night-shift work has been reported to adjust the sleep duration, as well as alertness, in real night-shift operations on Norwegian oil production platforms.[50]

The effects of night-shift work on sleep can linger for decades. For example, one study concluded that the amount of sleep and related sleep disturbances in present-day workers were positively related to their previous experience of night-shift work.[51] Analysis of sleep clinic patients revealed an overrepresentation of former night-shift workers among the patient population.[52] In one very interesting twin study in which one twin worked night shifts, whereas the other did not, the twin who worked night shifts displayed deteriorated sleep quality even post-retirement![53]

For many people, the night-shift-induced sleep disorders can further contribute to decreased quality of life, impaired cognitive function, and increased risk of accidents.

As noted, several studies have reported that night-shift workers experience shorter sleep duration and reduced sleep efficiency compared to those who work during the day.[48] Night-shift workers often experience fragmented sleep, characterized by multiple awakenings during the day, which usually leads to decreased overall sleep quality. The disruption of sleep architecture, including decreased REM sleep and altered distribution of sleep stages, further contributes to compromised sleep quality among night-shift workers.

Night-shift workers are much more prone to developing sleep disorders compared to those who work during the day. As noted, SWSD is a common consequence, characterized by difficulties falling asleep or staying asleep, as well as excessive sleepiness during work hours. Studies have indicated that the prevalence of SWSD among night-shift workers ranges from 10% to 40%.[54] Additionally, the risk of developing other sleep disorders, such as sleep apnea and restless leg syndrome, is also higher among night-shift workers.

Inadequate sleep due to night-shift work has been associated with a wide range of physical and mental health problems. Night-shift workers commonly experience higher levels of stress, depression, and anxiety due to the disruption of their sleep-wake patterns and social life.[55] Night-shift work poses challenges in maintaining social relationships and balancing work and family responsibilities. The inverted schedule often leads to social isolation, as night-shift workers are awake and working while their family and friends are asleep or engaged in daytime activities. This isolation can result in feelings of alienation and reduced social support. Night-shift work also affects family dynamics and can strain relationships, especially when it comes to childcare responsibilities and limited quality time with family members.[56]

To mitigate the negative effects of night-shift work on sleep, various coping strategies and interventions can be implemented. Adopting good sleep hygiene practices, such as maintaining a regular sleep schedule, creating a conducive sleep environment, and practicing relaxation techniques, can improve sleep quality.[57] Light exposure management, including the use of bright light during the night shift and minimizing exposure to bright light during the daytime, can help realign our circadian rhythms.[58] Additionally, strategic napping has been shown to enhance alertness and overall well-being among night-shift workers.[59]

Night-shift work disrupts the natural sleep-wake cycle and poses significant challenges to sleep quality and overall health. The effects of night-shift work on sleep can lead to a range of negative consequences, including sleep disorders, impaired cognitive function, and increased risk of physical and mental health problems. Night-shift work is likely a permanent feature of modern society. By implementing appropriate mitigation strategies, such as optimizing the work environment, educating workers on sleep hygiene, and implementing supportive policies, employers can help minimize the detrimental effects of night-shift work on sleep and promote the well-being of their employees.

The extent to which the negative consequences of night-shift work reflect impaired sleep, disrupted circadian rhythms, or an interaction of both factors remains unknown. Studies on nocturnal mice are useful because the effects of light at night on sleep can be separated from the effects of light at night on circadian clocks. Using mice, many of the negative outcomes of light at night have been reported to reflect disrupted circadian clocks independent from sleep disruptions.[60]

Nonetheless, clever studies of disrupted circadian rhythms in mice have revealed how these disruptions influence sleep. One method to examine circadian rhythms is through the use of T-cycles. T-cycles are typically light-dark cycles that are not 24 hours. For example, mice may be exposed to T-cycles of 22 hours (11 hours of light and 11 hours of dark) which allows the animals to synchronize (entrain), but be exposed to light at times during the internal night time. When the period of the light-dark cycle is close to the period of the biological clock (i.e., ~24 hours), entrainment remains possible. However, more extreme T-cycles (e.g. <20 or >28 hours) generally prevent entrainment, allowing the animals to "break free" from the light/dark conditions and display a period of about 24 hours. In a study of mice, electroencephalography (EEG; a measure of brain cell activity), locomotor activity, and core body temperature were recorded under a T-cycle of 20 hours (10 hours of light and 10 hours of dark).[61] As the circadian rhythms became misaligned with the environmental 20-hour T-cycle, the rhythmic distribution of sleep was abolished. The typical distribution of NREM sleep across the day was reinstated when the mice were realigned, suggesting a long-term effect of T-cycles on sleep intensity. These data suggest that circadian misalignment results in long-term effects on sleep, which may have important behavioral consequences for the animals.[61] The extent to which these results suggest that mouse studies might be used to understand the long-term consequences of night-shift work on sleep remains to be determined.

References

1. Fifel K, Videnovic A. 2021. Circadian and sleep dysfunctions in neurodegenerative disorders—an update. *Frontiers in Neuroscience*, 14:1471.

2. Colwell CS. 2021. Defining circadian disruption in neurodegenerative disorders. *Journal of Clinical Investigation*, 131(19). https://doi.org/10.1172/JCI148288.

3. Walker M. 2017. *Why We Sleep: Unlocking the Power of Sleep and Dreams*. Scribner, New York.

4. Siegel JM. 2005. Clues to the functions of mammalian sleep. *Nature*, 437(7063):1264–1271.

5. Cirelli C, Tononi G. 2008. Is sleep essential? *PLoS Biology*, 6(8): e216.

6. Vyazovskiy VV, Delogu A. 2014. The function of sleep: a bottom-up perspective. *Annual Review of Physiology*, 76: 235–259.

7. Zepelin H. 1989. Evolution of sleep: phylogenetic and functional perspectives. *Sleep*, 12(5): 447–471.

8. Czeisler CA. 2013. Perspective: casting light on sleep deficiency. *Nature*, 497: S13.

9. Goel N, Basner M. 2018. Circadian rhythms, sleep deprivation, and human performance. *Progress in Molecular Biology and Translational Science*, 160: 279–297.

10. Cingi C, Emre IE, Muluk NB. 2018. Jetlag related sleep problems and their management: a review. *Travel Medicine and Infectious Disease*, 24: 59–64.

11. Suhner A, Schlagenhauf P, Johnson R, Tschopp A, Steffen R, Hatz C. 2019. Comparative study to determine the optimal melatonin dosage

form for the alleviation of jet lag. *Chronobiology International*, 36(9): 1200–1211.

12. Cho YW, O'Donnell J, Kim H, Park SY, Lee J. 2017. Sleep disruption and depression in veterans with obstructive sleep apnea: the role of negative cognitions. *Journal of Clinical Sleep Medicine*, 13(3): 425–432.

13. Fowler PM, Knez W, Crowcroft S, Mendham AE, Miller J, Sargent C, Halson S, Duffield R. 2017. Greater effect of east versus west travel on jet lag, sleep, and team sport performance. *Medicine and Science in Sports and Exercise*, 49: 2548–2561.

14. Lemmer B, Kern RI, Nold G, Lohrer H. 2002. Jet lag in athletes after eastward and westward time-zone transition. *Chronobiology International*, 19: 743–764.

15. Reilly T, Atkinson G, Budgett, R. 2001. Effect of low-dose temazepam on physiological variables and performance tests following a westerly flight across five time zones. *International Journal of Sports Medicine*, 22(3): 166–174.

16. Herxheimer A, Petrie KJ. 2002. Melatonin for the prevention and treatment of jet lag. *Cochrane Database of Systematic Reviews*, 2: CD001520.

17. Foster RG, Peirson SN, Wulff K, Winnebeck E, Vetter C, Roenneberg T. 2013. Sleep and circadian rhythm disruption in social jetlag and mental illness. *Progress in Molecular and Biological Translational Science*, 119: 325–346.

18. Caliandro R, Streng AA, van Kerkhof LWM, van der Horst GTJ, Chaves I. 2021. Social jetlag and related risks for human health: a timely review. *Nutrients*, 13(12): 4543.

19. Wittmann M., Dinich J., Merrow M., Roenneberg T. 2006. Social jetlag: misalignment of biological and social time. *Chronobiology International*, 23: 497–509.

20. Roenneberg T, Wirz-Justice A, Merrow M. 2003. Life between clocks: daily temporal patterns of human chronotypes. *Journal of Biological Rhythms*, 18: 80–90.

21. Parsons MJ, Moffitt TE, Gregory AM, Goldman-Mellor S, Nolan PM, Poulton R, Caspi A. 2015. Social jetlag, obesity and metabolic disorder: investigation in a cohort study. *International Journal of Obesity*, 39: 842–848.

22. Wittmann M, Dinich J, Merrow M, Roenneberg T. 2006. Social jet-lag: misalignment of biological and social time. *Chronobiology International*, 23: 497–509.

23. Henderson SEM, Brady EM, Robertson N. 2019. Associations between social jetlag and mental health in young people: a systematic review, *Chronobiology International*, 36: 1316–1333.

24. Foster RG, Peirson SN, Wulff K, Winnebeck E, Vetter C, Roenneberg T. 2013. Sleep and circadian rhythm disruption in social jetlag and mental illness. *Progress in Molecular Biology and Translational Science*, 119: 325–346.

25. Tubbs AS, Hendershot S, Ghani SB, Nadorff MR, Drapeau CW, Fernandez F-X, Perlis ML, Grandner MA. 2022. Social jetlag and other aspects of sleep are linked to non-suicidal self-injury among college students. *Archives of Suicide Research*, 10: 1080.

26. Navarro-Martínez R, Chover-Sierra E, Colomer-Pérez N, Vlachou E, Andriuseviciene V, Cauli O. 2020. Sleep quality and its association with substance abuse among university students. *Clinical Neurology and Neurosurgery*, 188: 105591.

27. Qu Y, Li T, Xie Y, Tao S, Yang Y, Zou L, Zhang D, Zhai S, Tao F, Wu X. 2023. Association of chronotype, social jetlag, sleep duration and depressive symptoms in Chinese college students. *Journal of Affective Disorders*, 320: 735–741.

28. Wesley KL, Cooper EH, Brinton JT, Meier M, Honaker S, Simon SL. 2023. A national survey of U.S. adolescent sleep duration, timing, and social jetlag during the COVID-19 pandemic. *Behavioral Sleep Medicine*, 21: 291–303.

29. Díaz-Morales JF, Escribano C. 2015. Social jetlag, academic achievement and cognitive performance: understanding gender/sex differences. *Chronobiology International*, 32: 822–831.

30. Blume C, Garbazza C, Spitschan M. 2019. Effects of light on human circadian rhythms, sleep and mood. *Somnologie* (Berl), 23: 147–156.

31. Wright KP, Mchill AW, Birks BR, et al. 2013. Entrainment of the human circadian clock to the natural light-dark cycle. *Current Biology*, 23: 1554–1558.

32. Viola AU, James LM, Schlangen LJ, et al. 2008. Blue-enriched white light in the workplace improves self-reported alertness, performance and

sleep quality. *Scandinavian Journal of Work and Environmental Health*, 34: 297–306.

33. Boubekri M, Cheung IN, Reid KJ, et al. 2014. Impact of windows and daylight exposure on overall health and sleep quality of office workers: A case-control pilot study. *Journal of Clinical Sleep Medicine*, 10: 603–611.

34. Figueiro MG, Steverson B, Heerwagen J, et al. 2017. The impact of daytime light exposures on sleep and mood in office workers. *Sleep Health*, 3: 204–215.

35. Wams EJ, Woelders T, Marring I, et al. 2017. Linking light exposure and subsequent sleep: a field polysomnography study in humans. *Sleep*, https://doi.org/10.1093/sleep/zsx165.

36. Rajaratnam SM, Arendt J. 2001. Health in a 24-h society. *Lancet* 358: 999–1005.

37. National Sleep Foundation. 2014. Sleep in America poll. National Sleep Foundation, Arlington, VA, 2014.

38. Mason IC, Grimaldi D, Reid KJ, Warlick CD, Malkani RG, Abbott SM, Zee PC. 2022. Light exposure during sleep impairs cardiometabolic function. *Proceedings of the National Academies of Science*, 119: e2113290119.

39. Cajochen C, Munch M, Kobialka S, et al. 2005. High sensitivity of human melatonin, alertness, thermoregulation, and heart rate to short wavelength light. *Journal of Clinical and Endocrinological Metabolism*, 90: 1311–1316.

40. Chang A-M, Aeschbach D, Duffy JF, et al. 2015. Evening use of light-emitting eReaders negatively affects sleep, circadian timing, and next-morning alertness. *Proceedings of the National Academies of Science*, 112: 1232–1237.

41. Schweizer A, Berchtold A, Barrense-Dias Y, et al. 2017. Adolescents with a smartphone sleep less than their peers. *European Journal of Pediatrics*, 176: 131–136.

42. Christensen MA, Bettencourt L, Kaye L, Moturu ST, Nguyen KT, Olgin JE, Pletcher MJ, Marcus GM. 2016. Direct measurements of smartphone screen-time: relationships with demographics and sleep. *PLoS One*, 11: e0165331.

43. Souman JL, Borra T, de Goijer I, et al. 2018. Spectral tuning of white light allows for strong reduction in melatonin suppression without

changing illumination level or color temperature. *Journal of Biological Rhythms*, 33: 420–431.

44. Allen AE, Hazelhoff EM, Martial FP, et al. 2018. Exploiting metamerism to regulate the impact of a visual display on alertness and melatonin suppression independent of visual appearance. *Sleep*, https://doi.org/10.1093/sleep/zsy100.

45. Costa G, Haus E. 2010. Shift work and health: current problems and preventive actions. *Safety and Health at Work*, 1: 112–123.

46. Åkerstedt T, Wright KP. 2009. Sleep loss and fatigue in shift work and shift work disorder. *Sleep Medicine Clinics*, 4: 257–271.

47. Folkard S. 2008. Do permanent night workers show circadian adjustment? A review based on the endogenous melatonin rhythm. *Chronobiology International*, 25: 215–224.

48. Bjorvatn B, Stangenes K, Oyane N, Forberg K, Lowden A, Holsten F, et al. 2006. Subjective and objective measures of adaptation and readaptation to night work on an oil rig in the North Sea. *Sleep*, 29: 821–829.

49. Dumont M, Montplaisi J, Infante-Rivard C. 1988. Insomnia symptoms in nurses with former permanent nightwork experience. In: *Sleep '86*, WP Koella, F Obal, H Schultz, P Visser (Eds.), pp. 405–406. Gustav Fischer Verlag, Stuttgart.

50. Dement WC, Hall J, Walsh JK. 2003. Tiredness versus sleepiness: semantics or a target for public education? *Sleep*, 26: 485–486.

51. Ingre M, Akerstedt T. 2004. Effect of accumulated night work during the working lifetime, on subjective health and sleep in monozygotic twins. *Journal of Sleep Research*, 13: 45–48.

52. Barger LK, Ogeil RP, Drake CL, O'Brien CS, Ng KT, Rajaratnam SM, Zhou X. 2019. Predictors of key outcomes of sleep and sleep disorders in the United States: an extension of the national comorbidity survey replication (NCS-R). *Sleep*, 42: 253.

53. Gerber M, Hartmann T, Brand S, Holsboer-Trachsler E, Pühse U. 2010. The relationship between shift work, perceived stress, sleep and health in Swiss police officers, *Journal of Criminal Justice*, 38: 1167–1175.

54. Härmä M. 2019. Workhours in relation to work stress, recovery and health. *Scandinavian Journal of Work, Environment and Health*, 45: 321–326.

55. Scott LD, Hwang WT. 2012. Effects of nursing interventions promoting sleep on sleep quality in adults in health care settings: a systematic review. *Sleep Medicine Reviews*, 16: 389–399.

56. Gumenyuk V, Roth T, Drake CL. 2012. Circadian phase, sleepiness, and light exposure assessment in night workers with and without shift work disorder. *Chronobiology International*, 29: 928–936.
57. Kecklund G, Axelsson J, Åkerstedt T. 2016. Sleepiness and performance in night work: the role of countermeasures. *Sleep Medicine Reviews*, 26: 15–22.
58. Borniger JC, Weil ZM, Zhang N, Nelson RJ. 2013. Dim light at night does not disrupt timing or quality of sleep in mice. *Chronobiology International*, 30: 1016–1023.
59. Hasan S, Foster RG, Vyazovskiy VV, Peirson SN. 2018. Effects of circadian misalignment on sleep in mice. *Scientific Reports*, 8: 15343.
60. Cingi C, Emre IE, Muluk NB. 2018. Jetlag related sleep problems and their management: A review. *Travel Medicine and Infectious Disease*, 24: 59–64.
61. Nedergaard M, Goldman SA. 2020. Glymphatic failure as a final common pathway to dementia. *Science*, 370: 50–56.

5

LIGHT, COGNITION, AND MEMORY

Cognition comprises mental processes associated with acquiring, retrieving, and understanding information through thought, experience, and sensory input. Circadian rhythms in human cognition, primarily learning and memory, have been known to exist for decades. Early studies examining the relationship between time of day and learning and memory revealed intriguing diurnal patterns.[1] Research conducted in the 1960s and 1970s indicated that cognitive performance, including memory recall and learning abilities, exhibit significant fluctuations throughout the day. Studies by Kleitman[2] and his colleagues demonstrated better memory retention and recall during morning hours compared to the afternoon or evening, suggesting that circadian rhythms play a crucial role in cognitive processes.

Learning can be defined as a process that expresses itself as an adaptive change in behavior in response to experience.

Dark Matters. Randy J. Nelson, Oxford University Press. © Oxford University Press (2025).
DOI: 10.1093/9780197639979.003.0005

The stages of learning include acquisition (taking in information), consolidation (transforming experiences into memories), retrieval (recalling information), and extinction (forgetting information). Memory—the encoding, storage, and retrieval (or failure of retrieval, aka forgetting) of information about past experience—is necessary for learning to take place.

A memory system is required for adaptive changes in future behaviors to result from experience. Current situations must be compared with prior events that are in some way encoded in memory. Memory has been categorized into several types, to aid in both the description and the study of neural mechanisms underlying memory. However, even with these descriptive labels, there is often disagreement among experts about their definitions and what those definitions reveal about underlying mechanisms. For example, memory can be divided into short-term and long-term memory. Short-term memories last for seconds to minutes. For example, when you look up a zip code for someone, you usually can retain it sufficiently long to input the number. However, if you are distracted for a moment, you may have to look it up again if you did not rehearse the zip code to yourself several times. Such rehearsing is the best way to move items from short-term into long-term memory. Long-term memory persists for days, weeks, or years. Indeed, our long-term memory appears to have no upper limit in capacity or retention, whereas our short-term memory seems able to handle about seven items, give or take a few items—about the number of digits in a telephone number—for up to an hour.

Long-term memory can be divided into several categories as well. For instance, it can be divided into procedural (implicit) memory and declarative (explicit) memory. Procedural memory can be considered your memory for "knowing how." There are three types of procedural memory, but so-called skill learning is a type of procedural memory that is probably well-known to you. For instance, recalling how to ride a bicycle or play a song on a guitar are examples of procedural memory for skill learning. You may also be familiar with conditioning (e.g., Pavlov's dogs), which is another type of procedural memory. Declarative memory is your memory for "knowing what," that is, for knowing facts. Declarative memory comprises both semantic and episodic memory. Semantic memory is your general knowledge of facts and events, whereas episodic memory is your memory of personal events (episodes), such as your 30th birthday party. Declarative memory has also been called verbal memory in humans; procedural memories are generally nonverbal. Declarative memories are generally acquired and lost relatively easily, whereas procedural memories, such as knowing how to swim, require longer to establish but, once learned, are retained for long periods of time.

A final way to categorize memory is by dividing it into working memory and reference memory. Working memory is similar to declarative and short-term memory in that it typically involves short-term memory for information that changes on a regular basis. Working memory differs from reference memory, which generally refers to associations or discriminations requiring repetitious learning, as in learning the rules associated with driving a car.

The ability to learn, store, and retrieve information varies across the day. Peak learning and memory performance are typically observed during the daytime among diurnal species (e.g., humans) and during the nighttime among nocturnal species (e.g., laboratory mice and rats).[3,4] For example, rats display typical memory responses when trained early in the dark (active) phase, but memory acquisition was impaired if the rats were trained early in the light (inactive) phase.[5] Similarly, the ability to recall newly acquired information shows a 24-hour rhythm relative to the time information was acquired[4,6]; shifts of the light-dark cycle either before or after the learning occurred disrupt future memory. Removing the central brain clock (the suprachiasmatic nuclei, or SCN) in rats eliminates the 24 hour cycle of enhanced memory recall after learning a passive avoidance task.[6] Passive avoidance learning refers to avoiding a situation altogether that involves an aversive stimulus. For example, we might choose to simply not go out at night so that we do not venture into dark alleys. In this case, we are not performing an action, but rather avoiding taking action to avoid a negative consequence. In rodents, a common passive avoidance task is a two-chamber device in which one chamber is brightly lit, whereas the other chamber is dark and preferred. If there is a mild foot shock experienced in the dark box, the mouse ought to avoid going into that chamber until it "forgets." Typical clock function is required for learning and memory. As we will see later in this chapter, studies in mice with various clock genes deleted display aberrant clock function and impaired learning and memory. Also, if one examines the circadian clock gene expression in the hippocampus and other brain structures associated with learning and memory, then a typical rhythmic pattern is observed.

Jet Lag

Jet lag symptoms resolve quickly, especially when fewer than six time zones are crossed, and are mainly an annoyance for casual travelers.[7] However, for people who travel frequently, including flight personnel, the effects of chronic jet lag can linger.

Chronic jet lag can have serious effects on memory function. In one study, flight crews who worked on transmeridian (crossing multiple time zones) flights were compared with ground crews.[·] The flight crews displayed increased salivary cortisol concentrations relative to the ground crews; cortisol is a metabolic hormone with a strong circadian rhythm that is further increased in response to a stressor. Importantly, the flight crews also displayed impaired memory. To rule out the stress of flying, a follow-up study was conducted on two sets of flight crews. One flight crew was designated as "short-recovery," as they had less than 5 days to recover from international flights (>7 time zones), whereas the other flight crew was designated as "long-recovery" and had more than 15 days to recover from such flights. The flight crews were similar in most other respects. Five female flight attendants in each group received a functional MRI scan of their brains. The flight attendants who were working under the short-recovery schedule had smaller hippocampal volumes than individuals working the long-recovery schedule (Figure 5.1). The hippocampus is the part of the brain where many memory processes occur and negative feedback (turning off) from the stress response is accomplished; having a smaller hippocampus has been associated with cognitive deficits in a wide range of studies. Likewise, the short-recovery flight attendants also displayed

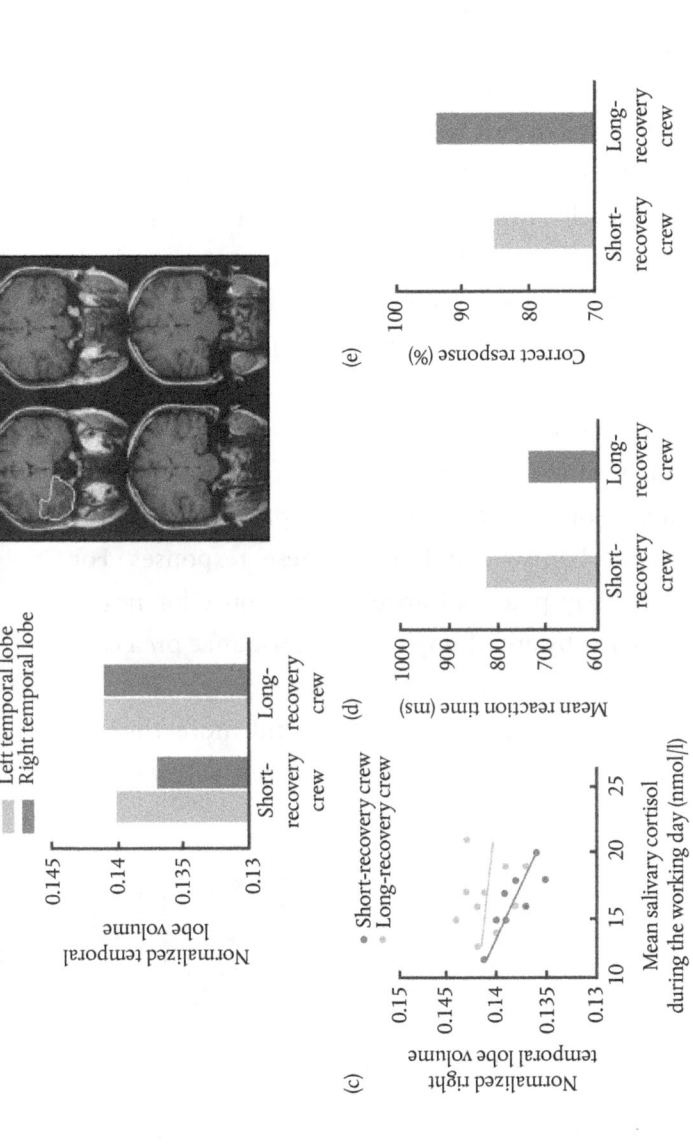

Figure 5.1 Effects of jet lag on brain and behavior. Chronic jet lag (a) decreased the volume of the hippocampus, a structure involved in learning and memory and outlined in (b), and increased cortisol values (c) and memory deficits (d and e). Source data from Cho K. (2001). Chronic "jet lag" produces temporal lobe atrophy and spatial cognitive deficits. *Nature Neuroscience*, 4: 567–568, Cho K, et al. (2000). Chronic jet lag produces cognitive deficits. *Journal of Neuroscience*, 20: 1–5. Reproduced with permission from Nelson, R.J. and Kriegsfeld L.J. (2022). *An Introduction to Behavioral Endocrinology*, Sixth Edition. Oxford, UK: Oxford University Press.

elevated cortisol and significant cognitive impairments, in addition to having a smaller hippocampus.'

Most of the cognitive tests given to the flight attendants involved assessments of working memory, primarily delayed match-to-sample tests. In this type of memory test, individuals are presented with a sample stimulus (e.g., a triangle). After a brief delay, the sample stimulus is shown again beside a novel alternative (e.g., a square) and the task is to remember which stimulus was previously seen. Typically, these stimuli are more complex than a triangle or square (Figure 5.2). In any case, the results of these studies suggest that chronic jet lag has significant cognitive consequences.

Animal models of jet lag also have demonstrated significant impairments of cognition and have provided insights into the brain mechanisms underlying these responses. For instance, twice-weekly phase advances of six hours for nearly a month provoked substantial impaired performance on a conditioned place preference test; in this memory test animals are assessed to determine whether they spend more time in a compartment where they previously received a reward or

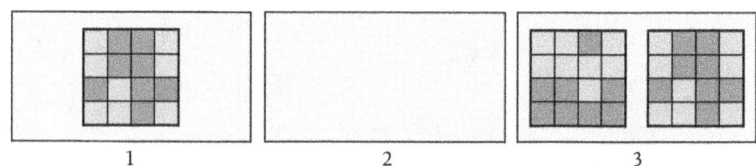

Figure 5.2 Example of delayed match to sample. In this example of a common memory test, participants would see screen 1 for a few seconds, then exposed to a blank screen 2 for some amount of time. After a predetermined delay, they would then see box 3 and have to choose the correct image (the one of the right in this case).

elsewhere. The impaired memory performance persisted for a month after the jet lag time shifts stopped.[8] The persisting effects of jet lag on learning and memory in rodents (and people) may be explained by reduced neurogenesis, a process of adding new nerve cells (neurons) to memory circuits. Although the functional role of adult neurogenesis remains somewhat unknown in adult mammals, there is increasing evidence that adult neurogenesis has an important role in learning and memory.[8,9] New neurons are made in adults in a section of the brain (subgranular zone of the dentate gyrus) associated with the hippocampus, a brain region important for several types of memory. Disrupting circadian rhythms impedes neurogenesis and negatively influences hippocampus-dependent memory.[8] As mentioned, phase advance (travel west) is easier than phase delay (travel east) in terms of jet lag recovery. It is interesting that reduced neurogenesis is "direction-dependent." Rats that experience six-hour phase advances displayed more significant reductions in neurogenesis than rats experiencing six-hour phase delays.[10] If these results pertain to human flight crews, then the impaired memory may reflect the reduction in the formation of new nerve cells or connections among nerve cells.

In addition to altered neurogenesis, changes in neuronal spine density and dendrite complexity are correlated with the formation of new connections (synapses) between nerve cells during memory formation.[11] Dendritic spines are small physical protrusions that allow communication between neurons. There are daily oscillations in the density of dendritic spines in the hippocampus of rodents; increased spine density is observed during the active part of the day.[12,13] Similarly, rats undergoing weekly six-hour phase advances for eight weeks have decreased dendritic complexity on immature neurons.[10]

Even a 1-hour, unavoidable phase shift can be significant for some people. In many industrialized nations, including the United States, most of the population undergoes a one-hour phase shift twice a year with the change from Standard Time to Daylight Savings Time. It usually requires several days for people to make the adjustment to this biennial time shift in their circadian systems.[14,15] In the United States, the rate of traffic accidents increases significantly in the week following the time change.[14,15] Whether this statistic reflects an impairment of attention and psychomotor coordination because of alterations in our circadian systems or whether it is due to some other indirect factor, such as missed appointments because of incorrectly set watches, requires further research.

Social Jet Lag

One way that social jet lag[16] impairs learning and memory is by disrupting the consolidation of new information into long-term memory. Sleep plays a crucial role in memory consolidation, particularly for declarative memory, which involves the encoding and retrieval of facts and events. Several studies have reported that insufficient or disrupted sleep negatively affects memory performance.[17]

For example, one study investigated the relationship between social jet lag and cognitive function in young adults. In this study, individuals with higher social jet lag scores performed worse on memory tasks compared to those with lower social jet lag scores.[18] Another study examined the effects of social jet lag on sleep quality and cognitive function in adolescents. Individuals in this study with higher social jet lag scores reported poorer sleep quality and demonstrated lower cognitive performance, including reduced memory consolidation, compared to those with low social jet lag scores.[19]

Social jet lag can impair attention and cognitive functioning, which are crucial for effective learning. Sleep deprivation resulting from social jet lag can lead to increased daytime sleepiness, decreased alertness, and reduced cognitive performance.[20] These cognitive deficits can impede information processing, attention allocation, and the ability to acquire new knowledge efficiently.

Another study focused on the effects of sleep deprivation on cognitive performance. Participants who were sleep deprived displayed significant impairments in attention and working memory tasks compared to well-rested individuals.[21] These findings indicate that social jet lag, which disrupts sleep patterns and reduces sleep duration, can compromise the brain processes associated with paying attention, which is necessary for effective learning.

An increasing awareness of social jet lag, sleep deprivation, and school start times has recently occurred. Adolescents experience a shift in their circadian rhythms to evening chronotypes, resulting in a natural tendency to stay up later and wake up later. This sleep phase delay often conflicts with early school start times, leading to inadequate sleep and poor academic performance.[22-24] It is tempting to simply assert that teenagers lack discipline in maintaining good sleep hygiene, but during adolescence, there is a delay in the onset of melatonin secretion; this contributes to difficulty falling asleep early in the evening, and making it challenging for adolescents to align their sleep schedules with early school start times.[25]

Cognitive functions crucial for academic performance, including attention, memory, and problem-solving, are negatively affected by sleep deprivation. Many studies have established that sleep deprivation in adolescents is associated with decreased attention, reduced working memory capacity, and impaired executive functions.[26,27] Early school start times often lead to inadequate sleep duration and fragmented sleep.[28] In contrast, it has been reported that delayed school start times are associated with improved attention, reduced daytime sleepiness, and increased academic performance among adolescents.[29,30] Indeed, the later school start times were associated with significant improvements in academic performance, including higher grades and standardized test scores. In another study, adequate sleep resulting from later start times improves students' alertness, attention, and cognitive function.[31]

Implementing later start times poses logistical challenges for schools, transportation systems, and extracurricular activities. Careful planning and coordination can address these concerns effectively.[32] Implementing later start times gradually, through pilot programs, can help identify and address potential challenges before full-scale implementation. Such pilot studies can collect data on outcomes, logistics, and student well-being to inform decision-making. There have been numerous studies that demonstrated positive effects of later start times on academic performance. For example, one study in a large urban school district in Minnesota reported significant improvements in attendance rates, graduation rates, and academic performance when later start times were implemented.[33] The Seattle School

District in Washington shifted high school start times from 7:50 a.m. to 8:45 a.m. This later start time resulted in improved academic performance, and well-being, among the students.[34,35] Importantly, the long-term implications of implementing later start times include benefits for students' lifelong learning, sleep habits, and long-term academic and subsequent career success.[36]

Based on the available evidence, parents, policymakers, educators, and health professionals should advocate for policy changes that prioritize the implementation of later start times. The American Academy of Pediatrics[37] issued a policy statement recommending middle and high schools start no earlier than 8:30 a.m. to ensure optimal sleep for students. Adolescents phase delay their sleep time naturally via delayed melatonin secretion onset; social jet lag often shifts sleep time later, especially on the weekends. Starting school later to accommodate the effects of disrupted circadian rhythms on cognition is a rational approach to optimizing the success of our teenagers. No studies have been reported suggesting that social jet lag influences brain structures similarly to the study results depicted in Figure 5.2.

Exposure to Dim Lighting During the Day

The effects of bright light during the day in the regulation of cognitive processes have been reported for many species. Typically, bright illumination evokes enhanced cognitive performance. In terms of school performance, brighter light in the classroom has been reported to increase cognitive performance of elementary school students in math and reading.[38] This is a critical point. Since the energy crisis in the 1970s, schools have been designed and built with narrow windows that allow less heated or cooled air to escape the buildings, but these energy efficient windows also do not allow much sunlight to enter the schools. Indeed, children's classrooms are designed to only have 300 lux of light during the day.[39] This is very likely an insufficient light intensity to maintain proper circadian organization. Adults also benefit from bright offices. Increasing the illumination of offices improves the cognitive performance of adult workers.[40,41]

Bright light modulates human attention and executive functions involved in cognitive processing. Daytime bright light promotes alertness, which is essential for optimal cognitive function, and also reduces sleepiness and daytime fatigue compared to people maintained in dim daytime light conditions.[42] Brain imaging studies revealed that subcortical (deeper) regions that support alertness/arousal are activated by daytime bright light exposure prior to activation of cortical areas involved in cognition.[43]

Similarly, fMRI imaging revealed that daytime exposure to blue (470 nm) monochromatic light during a complicated auditory working memory test influenced regional brain responses.[44] Blue light typically enhanced brain responses in the frontal and parietal cortices implicated in working memory, and in the thalamus involved in arousal, which modulates cognition. The results of this study indicate that short wavelength (blue) light can affect cognitive functions almost instantaneously and suggest that these effects are mediated by the melanopsin-based photoreceptor system associated with providing light stimuli to the primary circadian clock. The blue-enriched light used in this study had a 460 nm wavelength.[44]

Living in chronically dim light during the day is especially common for many individuals suffering from neurodegenerative disorders, such as Alzheimer's disease (AD). AD is the most common form of dementia. Although aging is an important risk factor for AD, it is possible that disrupted circadian rhythms may also be an important, but largely unrecognized, risk factor. Indeed, bright light therapy has been reported to reduce cognitive deterioration in mild/early-stage AD.[45]

Disrupted circadian rhythms and neurodegeneration interact in a vicious cycle that accelerates the progression of AD. For instance, circadian rhythms seem to degrade during aging.[46] Changes associated with aging are commonly linked to variability, both intra-daily and inter-daily variability. For example, elderly folks may sporadically show a dissociation between their body temperature and locomotor activity. Importantly, aging interferes with the ability to synchronize to changing light-dark cycles such as traveling across multiple time zones.[46] Animal studies reveal that inactivating a key circadian clock gene, $Bmal1$, evokes brain neuroinflammation and neurodegeneration, and also impedes learning and memory![47]

Amyloid-beta (Aβ) is a relatively large protein that is found in nerve cell membranes and typically plays an important role in cell growth and repair. However, as we age this protein may become corrupted, then aggregate to form the extracellular plaques characteristic of AD. These plaques can destroy nerve cells, which leads to impaired cognition. In common with most biological functions, the dynamics of Aβ are controlled by circadian clocks. Conversely, the buildup of plaque interferes with clock function, at least in mice.[47] Both the neuropathology and the disrupted circadian rhythms take years to develop or notice. However, it might be possible to observe the changes in circadian rhythms, and ameliorate the changes to help with AD.

Again, typical age-related changes in circadian rhythms include reduced amplitude (for example, the peak and lowest body temperature difference across the day may decline), decreased responsiveness to synchronizing factors (even relatively bright light is not effective at synchronizing the internal biological clocks), and reduced stability of the timing mechanisms.[47] Additionally, daily sleep-wake rhythms are more fragmented in AD patients; AD patients typically display increased nocturnal awakenings and daytime naps.[48] Atypical sleep-wake rhythms are often observed many years prior to the emergence of cognitive deficits and tend to worsen as AD progresses.[49] Thus, recognition of these patterns may allow earlier diagnosis and onset of treatments for AD. One important role of sleep is to allow for brain clearance of various waste products including Aβ. Plaques in the brain are formed when pieces of Aβ clump together. Aβ comes from a larger protein found in the fatty membrane surrounding nerve cells. Amyloid-beta is chemically "sticky" and gradually builds up into plaques; this buildup of Aβ as amyloid plaques in the brain is a hallmark of AD. The loss of consolidated sleep likely contributes to amyloid plaque burden, worsening AD symptoms.

Although the precise mechanisms by which aging impairs circadian rhythms remain unspecified, there are several nonexclusive hypotheses.[47] First, eye function wanes during aging. The cornea often becomes increasingly opaque (cloudy) during aging, reducing the light information that reaches the retina, and in turn, the primary circadian clock. Second, there is evidence that the primary clock itself displays less robust rhythms in aged mice. Aging also changes the quality and quantity of neurotransmitters (the chemical messengers that travel between nerve cells) both within the primary clock and in the signals from the clock to the rest of the body.[46] Other brain cells such as glia are also influenced by aging, often becoming hyperactive. Glial cells are brain immune cells that are crucial in providing support and protecting the nerve cells, maintaining homeostasis, cleaning up cellular debris, and forming myelin (the fatty sheath around components of the nerve cells that allows them to communicate quickly and efficiently with other nerve cells). In other words, in a healthy brain, glia work to provide an optimal environment for neurons to function. These changes in neurotransmitters and glial function observed during typical aging seem especially true of AD patients' postmortem brains.

Elderly folks who live in institutional housing rarely experience exposure to bright daytime light. Low intensity daytime light could increase the risk of reduced SCN clock gene expression rhythms and subsequent flattened circadian rhythms in AD patients as it does in rodents.[47] In one intriguing study of residents of an assisted living facility, increased exposure to morning light was associated with less fragmented and more stable rest–activity rhythms.[50] Examination of 53 studies on this topic via a meta-analysis concluded that bright daytime light exposure was an effective therapy for circadian rhythm sleep disorders in AD.[51] The extent to which restoring sleep will ameliorate the symptoms of AD remains to be determined.[52] However, this possibility is exciting as it is much easier to modify environmental light exposure to reduce AD risk or symptom expression compared to modifying genetic risk factors.

The mechanisms underlying impaired cognition under dim light during the day were investigated in a study of diurnal rats. Unlike most species of mice and rats, these Nile grass rats are active during the day, similarly to people, chipmunks, and squirrels. Dim light during the day impairs spatial memory in these diurnal creatures. Providing them with bright light (1,000 lux) during the day reversed the learning deficits.[53] The results suggest that daytime light intensities influence cognitive function of grass rats similarly to people; that is, bright light is beneficial compared to dim light for cognitive performance. In addition to the altered cognitive abilities, grass rats in dim daytime light intensities also displayed reduced expression of the gene coding for brain-derived neurotrophic factor (BDNF) in the hippocampus. BDNF plays an important role in neuronal survival and growth, modulates neurotransmitter function, and is involved in neuronal plasticity, which is essential for learning and memory. The changes in BDNF gene expression were observed most notably in the CA1 subregion of the hippocampus, which is specifically involved in spatial memory. There were also physical changes in this portion of the hippocampus as a result of exposure to dim light levels during the day. Dendritic spine density was reduced in the CA1 neurons.[53] Remarkably, when the animals were transferred to the bright daytime lighting condition for four weeks, the hippocampal BDNF and dendritic spine density significantly increased! Similar results were obtained in both sexes.[54] These results suggest not only that light intensity during the day affects cognitive performance but also that there is a structural mechanism by which this occurs. Importantly, this study suggests the good news that brain and cognitive behavior deficits caused by exposure to dim light during the day can be reversed by daytime exposure to bright light![53-55]

Exposure to Light at Night

Most studies of the effects of light at night have been conducted in the context of night-shift work (for more on this, see the next section). Although there are dozens of studies reporting that exposure to light at night negatively affects sleep, and other studies have demonstrated that disrupted sleep is associated with impaired cognition, few studies have reported the direct effects of light at night on cognitive function in humans. The role of disrupted circadian rhythms on sleep and downstream effects on cognitive function are discussed in Chapter 4.

However, there have been some recent reports of impaired cognitive function in the elderly and disrupted circadian rhythms by light at night. Mild cognitive impairment is a transitional stage between typical aging and Alzheimer's disease/dementia in the elderly. A multicity study among veterans in China examined the relationship between outdoor light at night exposure and the incidence of mild cognitive impairment.[56] The results of this study revealed that exposure to bright outdoor light at night was associated with higher risk of mild cognitive impairment, such as mild memory loss and reduced ability to learn new information. These results suggest, in turn, that reducing exposure to light at night might reduce or prevent mild cognitive impairment.[56]

However, there have been dozens of experimental studies published over the past two decades, primarily in nocturnal rodents, showing that light at night disrupts circadian rhythms, and alters cognitive function. For instance, disruption of circadian rhythms by chronic exposure to dim light at night reduces hippocampal CA1 dendritic spine density and dendritic arborization in rodents.[57] Long-term exposure to light at night also decreases dendritic arborization in the brains of the diurnal Nile grass rats,[58] confirming similar effects in diurnal and nocturnal rodents. Dendritic spines are the connectors between nerve cells and it is generally accepted that reduced numbers of these spines and reduction of the formation of tree-like ultrastructures (arborization) leads to reduced nerve cell function. Other studies of the effects of dim blue light exposure revealed increased stress hormone concentrations and elevated neuroinflammation of the hippocampus, which caused neuronal loss and synaptic dysfunction. In terms of cognitive function, exposure to dim blue light at night disrupted circadian rhythms, which lead to spatial learning and memory dysfunction in mice.[59]

Night-Shift Work

Long-time night-shift workers have also been observed to display significant impairments in cognitive function. For example, both reaction time and short-term memory are worse among night-shift workers as compared to their day shift counterparts.[60-62] It is thus perhaps not surprising that major industrial accidents often occur during the night shift.[63] These industrial accidents include the tragedy at the Union Carbide chemical plant in Bhopal, India; the nuclear reactor meltdowns at Three Mile Island and Chernobyl; the grounding of the Exxon Valdez oil tanker; and the explosion and massive oil leak of the Deepwater Horizon rig in the Gulf of Mexico.

Likewise, sleep-deprived workers are 70% more likely to be involved in work-related accidents according to the US Institute of Medicine Committee on Sleep Medicine and Research.[64] Night-shift workers reporting disturbed sleep and excessive daytime sleepiness are almost twice as likely to die in a work-related accident.

Several studies have established the short-term effects of night-shift work on cognition, but a number of studies have reported rather compelling and disturbing long-term effects of night-shift work. For example, night-shift work has been estimated to reduce cognitive function similar to 6.5 years of age-related aging![65] Males, but not females, who currently work night shifts displayed lower cognitive performance than workers who had never worked night shifts. Memory and other assessed cognitive performance factors decreased as the duration of night-shift work increased. One positive aspect of this population was that the cognitive performance of workers who had stopped night-shift work more than four years ago slightly increased, suggesting that time could reverse the negative effects of night-shift work.[65] In summary, cognitive functioning tends to be impaired by a long-term exposure to night-shift work. How to manage disruption of circadian rhythms will be discussed in Chapter 8.

References

1. Folkard S. 1979. Time of day and level of processing. *Memory and Cognition*, 7: 247–252.
2. Kleitman N. 1963. *Sleep and Wakefulness*. University of Chicago Press, Chicago.

3. Evans, MDR, Kelley P, Kelley J. 2017. Identifying the best times for cognitive functioning using new methods: matching university times to undergraduate chronotypes. *Frontiers in Human Neuroscience*, 11: 188.

4. Chaudhury D, Colwell CS. 2002. Circadian modulation of learning and memory in fear-conditioned mice. *Behavioral Brain Research*, 133: 95–108.

5. Davies JA, Navaratnam V, Redfern PH. 1974. The effect of phase-shift on the passive avoidance response in rats and the modifying action of chlordiazepoxide. *British Journal of Pharmacology*, 51: 447–451.

6. Cain SW, McDonald RJ, Ralph MR. 2008. Time stamp in conditioned place avoidance can be set to different circadian phases. *Neurobiology, Learning and Memory*, 89: 591–594.

7. Office of Technology Assessment. 1991. Biological rhythms: Implications for the worker (OTA-BA-463). Government Printing Office, Washington, DC.

8. Gibson EM, Wang C, Tjho S, Khattar N, Kriegsfeld LJ. 2010. Experimental "jet lag" inhibits adult neurogenesis and produces long-term cognitive deficits in female hamsters. *PLoS One*, 5: e15267.

9. Cameron HA, Glover LR. 2015. Adult neurogenesis: beyond learning and memory. *Annual Review of Psychology*, 66: 53–81.

10. Horsey EA, Maletta T, Turner H, Cole C, Lehmann H, Fournier NM. 2019. Chronic jet lag simulation decreases hippocampal neurogenesis and enhances depressive behaviors and cognitive deficits in adult male rats. *Frontiers in Behavioral Neuroscience*, 13: 272.

11. Moeller JS, Kriegsfeld LK. 2023. Circadian rhythms and cognitive functioning. In: *Biological Implications of Circadian Disruption: A Modern Health Challenge*, LK Fonken, RJ Nelson (Eds.). Cambridge University Press, Cambridge, UK.

12. Ikeno T, Nelson RJ. 2015. Acute melatonin treatment alters dendritic morphology and circadian clock gene expression in the hippocampus of Siberian hamsters. *Hippocampus*, 25: 42–48.

13. Ikeno T, Weil ZM, Nelson RJ. 2013. Photoperiod affects the diurnal rhythm of hippocampal neuronal morphology of Siberian hamsters. *Chronobiology International*, 30: 1089–1100.

14. Monk TH. 1980. Traffic accident increases as a possible indicant of desynchronosis. *Chronobiologia*, 7: 527–529.

15. Monk TH, Aplin LC. 1980. Spring and autumn daylight savings time changes: studies of adjustment in sleep timings, mood, and efficiency. *Ergonomics*, 23: 167–178.

16. Foster RG, Peirson SN, Wulff K, Winnebeck E, Vetter C, Roenneberg T. 2013. Sleep and circadian rhythm disruption in social jetlag and mental illness. *Progress in Molecular Biology and Translational Science*, 119: 325–346.

17. Curcio G, Ferrara M, De Gennaro L. 2006. Sleep loss, learning capacity, and academic performance. *Sleep Medicine Reviews*, 10: 323–337.

18. Wittmann M, Dinich J, Merrow M, Roenneberg T. 2006. Social jetlag: misalignment of biological and social time. *Chronobiology International*, 23: 497–509.

19. Wilhelm I, Prehn-Kristensen A, Born J, Wilhelm FH. 2011. Sleep-dependent memory consolidation—what can be learnt from children? *Neuroscience and Biobehavioral Reviews*, 35: 1958–1971.

20. Roenneberg T, Wirz-Justice A, Merrow M. 2003. Life between clocks: daily temporal patterns of human chronotypes. *Journal of Biological Rhythms*, 18: 80–90.

21. Pilcher JJ, Lambert BJ, Huffcutt AI. 2000. Differential effects of permanent and rotating shifts on self-report sleep length: a meta-analytic review. *Sleep*, 23: 155–163.

22. Carskadon MA, Acebo C, Jenni OG. 2004. Regulation of adolescent sleep: implications for behavior. *Annals of the New York Academy of Sciences*, 1021: 276–291.

23. Carskadon MA, Acebo C. 2002. Regulation of sleepiness in adolescents: update, insights, and speculation. *Sleep*, 25: 606–614.

24. Carskadon MA, Wolfson AR. 1998. Adolescent sleep patterns, circadian timing, and sleepiness at a transition to early school days. *Sleep*, 21: 871–881.

25. Carskadon MA. 2011. Sleep in adolescents: the perfect storm. *Pediatric Clinics of North America*, 58: 637–647.

26. Dewald JF, Meijer AM, Oort FJ, Kerkhof GA, Bögels SM. 2010. The influence of sleep quality, sleep duration and sleepiness on school performance in children and adolescents: a meta-analytic review. *Sleep Medicine Reviews*, 14: 179–189.

27. Curcio G, Ferrara M, De Gennaro L. 2006. Sleep loss, learning capacity, and academic performance. *Sleep Medicine Reviews*, 10: 323–337.
28. Dewald JF, Meijer AM, Oort FJ, Kerkhof GA, Bögels SM. 2010. The influence of sleep quality, sleep duration and sleepiness on school performance in children and adolescents: a meta-analytic review. *Sleep Medicine Reviews*, 14: 179–189.
29. Wahlstrom K. 2017. Later school start times in the U.S.: an economic analysis (Research report). Arlington, VA: National Sleep Foundation.
30. Wahlstrom KL, Dretzke B, Gordon M, Peterson K, Edwards K. 2014. Examining the impact of later high school start times on the health and academic performance of high school students: a multi-site study. Center for Applied Research and Educational Improvement, University of Minnesota.
31. Vorona RD, Szklo-Coxe M, Wu A, Dubik M, Zhao Y. 2011. Dissimilar teen crash rates in two neighboring southeastern Virginia cities with different high school start times. *Journal of Clinical Sleep Medicine*, 7: 145–151.
32. Bruni O, Ball H, Owens J, Rajaratnam SM. 2016. Later school start times in adolescence: time for change. *Sleep*, 39: 687–688.
33. Owens J, Belon K, Moss P. 2010. Impact of delaying school start time on adolescent sleep, mood, and behavior. *Archives of Pediatrics and Adolescent Medicine*, 164: 608–614.
34. Wahlstrom K. 2017. Later school start times in the U.S.: an economic analysis (Research report). Arlington, VA: National Sleep Foundation.
35. Wahlstrom KL, Dretzke B, Gordon M, Peterson K, Edwards K. 2014. Examining the impact of later high school start times on the health and academic performance of high school students: a multi-site study. Center for Applied Research and Educational Improvement, University of Minnesota.
36. Owens J, Dalzell V, Winsler A. 2016. Sleep on it! Developing interventions to extend sleep duration. *Developmental Review*, 41: 1–22.
37. American Academy of Pediatrics. 2014. School start times for adolescents. *Pediatrics*, 134: 642–649.
38. Barkmann C, Wessolowski N, Schulte-Markwort M. 2012. Applicability and efficacy of variable light in schools. *Physiology and Behavior*, 105: 621–627.

39. National Research Council. 2007. *Green Schools: Attributes for Health and Learning*. Washington, DC: The National Academies Press. https://doi.org/10.17226/11756.

40. National Viola AU, James LM, Schlangen LJ, Dijk DJ. 2008. Blue-enriched white light in the workplace improves self-reported alertness, performance and sleep quality. *Scandinavian Journal of Work, Environment and Health*, 34: 297–306.

41. Mills P, Tomkins S, Schlangen L. 2007. The effect of high correlated colour temperature office lighting on employee well-being and work performance. *Journal of Circadian Rhythms*, 5: 2–10.

42. Ruger M, Gordijn MC, Beersma DG, de Vries B, Daan S. 2006. Time-of-day-dependent effects of bright light exposure on human psychophysiology: comparison of daytime and nighttime exposure. *American Journal of Physiology*, 290: R1413–R1420.

43. Vandewalle G, Balteau E, Phillips C, Degueldre C, Moreau V, Sterpenich V, Albouy G, Darsaud A, Desseilles M, Dang-Vu TT, Peigneux P, Luxen A, Dijk DJ, Maquet P. 2006. Daytime light exposure dynamically enhances brain responses. *Current Biology*, 16: 1616–1621.

44. Vandewalle G, Gais S, Schabus M, Balteau E, Carrier J, Darsaud A, Sterpenich V, Albouy G, Dijk DJ, Maquet P. 2007. Wavelength-dependent modulation of brain responses to a working memory task by daytime light exposure. *Cerebral Cortex*, 17: 2788–2795.

45. Riemersma-van der Lek RF, Swaab DF, Twisk J, Hol EM, Hoogendijk WJ, Van Someren EJ. 2008. Effect of bright light and melatonin on cognitive and noncognitive function in elderly residents of group care facilities: a randomized controlled trial. *Journal of American Medical Association*, 299: 2642–2655.

46. Duncan MJ. 2020. Interacting influences of aging and Alzheimer's disease on circadian rhythms. *European Journal of Neuroscience*, 51: 310–325.

47. Duncan MJ. 2023. Circadian rhythm disruption in aging and Alzheimer's disease. In: *Biological Implications of Circadian Disruption: A Modern Health Challenge*, LK Fonken, RJ Nelson (Eds.). Cambridge University Press, Cambridge, UK.

48. Bliwise DL, Mercaldo ND, Boever BF, Greer SA, Kukull WA. 2011. Sleep disturbance in dementia with Lewy bodies and Alzheimer's disease: a

multicenter analysis. *Dementia and Geriatric Cognitive Disorders*, 31: 239–246.

49. Bliwise DL, Hughes M, McMahon PM, Kutner N. 1995. Observed sleep/wakefulness and severity of dementia in an Alzheimer's disease special care unit. *Journal of Gerontology*, 50A: M303–M306.

50. Juda M, Liu-Ambrose T, Feldman F, Suvagau C, Mistlberger RE. 2020. Light in the senior home: effects of dynamic and individual light exposure on sleep, cognition, and well-being. *Clocks and Sleep*, 2: 557–576.

51. van Maanen A, Meijer AM, van der Heijden KB, Oort FJ. 2016. The effects of light therapy on sleep problems: a systematic review and meta-analysis. *Sleep Medicine Reviews*, 29: 52–62.

52. Li P, Gao L, Gaba A, Yu L, Cui L, Fan W, Lim ASP, Bennett DA, Buchman AS, Hu K. 2020. Circadian disturbances in Alzheimer's disease progression: a prospective observational cohort study of community-based older adults. *Lancet Healthy Longevity*, 1: e96–e105.

53. Soler JE, Robison AJ, Nunez AA, Yan L. 2017. Light modulates hippocampal function and spatial learning in a diurnal rodent species: a study using male Nile grass rat (*Arvicanthis niloticus*). *Hippocampus*, 2017: 1–12.

54. Soler JE, Stumpfig M, Tang YP, Robison AJ, Núñez AA, Yan L. 2019. Daytime light intensity modulates spatial learning and hippocampal plasticity in female Nile grass rats (*Arvicanthis niloticus*). *Neuroscience*, 404: 175–183.

55. Yan L, Lonstein JS, Nunez AA. 2019. Light as a modulator of emotion and cognition: lessons learned from studying a diurnal rodent. *Hormones and Behavior*, 111: 78–86.

56. Chen Y, Tan J, Liu Y, Dong G-H, Yang B-Y, Li N, Wang L, Chen G, Li S, Guo Y. 2022. Long-term exposure to outdoor light at night and mild cognitive impairment: a nationwide study in Chinese veterans. *Science of the Total Environment*, 847: 157441.

57. Bedrosian TA, Vaughn CA, Galan A, Daye G, Weil ZM, Nelson RJ. 2013. Nocturnal light exposure impairs affective responses in a wavelength-dependent manner. *Journal of Neuroscience*, 33: 13081–13087.

58. Fonken LK, Kitsmiller E, Smale L, Nelson RJ. 2012. Dim night-time light impairs cognition and provokes depressive-like responses in a diurnal rodent. *Journal of Biological Rhythms*, 27: 319–327.

59. Liu Q, Wang Z, Cao J, Dong Y, Chen Y. 2022. Dim blue light at night induces spatial memory impairment in mice by hippocampal neuroinflammation and oxidative stress. *Antioxidants*,11: 1218.

60. Marquié J-C, Tucker P, Folkard S, Gentil C, Ansiau D. 2015. Chronic effects of shift work on cognition: findings from the VISAT longitudinal study. *Occupational and Environmental Medicine*, 72: 258–264.

61. Meijman T, van der Meer O, van Dormolen M. 1993. The after-effects of night work on short-term memory performance. *Ergonomics*, 36: 37–42.

62. Rouch I, Wild P, Ansiau D, Marquié J-C. 2005. Shiftwork experience, age and cognitive performance. *Ergonomics*, 48: 1282–1293.

63. Fishbein AB, Knutson KL, Zee PC. 2021. Circadian disruption and human health. *Journal of Clinical Investigation*, 131: e148286.

64. Mitler MM, Carskadon MA, Czeisler CA, Dement WC, Dinges DF, Graeber RC. 1988. Catastrophes, sleep, and public policy: consensus report. *Sleep*, 11: 100–109.

65. Institute of Medicine (US) Committee on Sleep Medicine and Research. 2006. Functional and economic impact of sleep loss and sleep-related disorders. In: *Sleep Disorders and Sleep Deprivation: An Unmet Public Health Problem*, HR Colten, BM Altevogt (Eds.), pp. 137–172. National Academies Press (US).

6

LIGHT AND CANCER

Cancer. This is one of the most frightening words a patient can hear from their physician. Although substantial progress has been made in diagnosis, treatment, and outcomes for cancer patients, the psychological, physical, and financial tolls are substantial for many people. Currently, cancer treatment often involves a combination of surgical removal of the tumor and "poisoning" or directly attacking the tumor (i.e., chemotherapies, chimeric antigen receptor [CAR] T-cell therapy, or radiation). Cancers with the current highest five-year relative survival rates include melanoma (skin cancer) and Hodgkin lymphoma, as well as breast, prostate, testicular, cervical, and thyroid cancer. When treatment has been successful, patients are considered to be in "remission." The National Institutes of Health defines a cure as the point where the cancer is undetectable, no additional treatment is required, and the cancer is not expected to return. Physicians cannot be certain that cancer is cured after treatment, but many feel comfortable considering patients with five or more years of disease-free survival as cured. The longer a person is cancer-free, the better the chances are that the cancer will not

Dark Matters. Randy J. Nelson, Oxford University Press. © Oxford University Press (2025).
DOI: 10.1093/9780197639979.003.0006

return. I am a 40-year survivor of melanoma—it is unlikely that my skin cancer will return, but I'm still checked regularly for any signs of reoccurrence.

Before we proceed, let's define and describe cancer. It is not a unitary disease, but rather a collection of more than 100 different types of cancer that are typically named based on the tissue or organs from which the cancers originate. For example, breast cancer forms initially in the breast tissue, but can spread (metastasize) to the brain and other organs and cause significant problems that threaten function and even mortality. There are several categories of cancer that are based on the cell-types in which the cancer originates; these include carcinoma, sarcoma, leukemia, lymphoma, multiple myeloma, and others.

Carcinomas are the most common types of cancer; they are formed in epithelial cells, the columnar cells that line the inside and outside surfaces of our bodies. Common cancers of epithelial cells that produce fluids (e.g., glandular cells) include most cancers of the breast, colon, and prostate gland. Sarcomas form in bone and soft tissue, such as fat, muscle, lymph vessels and nodes, and fibrous tissue such as tendons and ligaments. Leukemias are cancers that form in bone marrow; unlike carcinomas and sarcomas, leukemias do not form solid tumors. Lymphoma is a cancer that originates in lymphocytes (B or T immune cells).

Typically, our cells grow and multiply via a process called cell division; after adulthood, new cells form only as our bodies need them. As cells are damaged, or simply grow old and less functional, they die and new cells take their place. However, in some cases, this orderly process breaks down, and abnormal or damaged cells grow and multiply when and where they should not. These cells may form tumors (i.e., clusters of dividing cells). Tumors can be cancerous (malignant) or not cancerous (benign). Cancerous tumors often invade nearby tissues and can also travel to distant places in the body to form new tumors (a process that is called metastasis). Benign tumors do not invade other tissues, but may grow to crowd adjoining tissue.

Cancer is a genetic disease. It is caused by changes to genes that control the way cells grow and divide to make new cells. Common genetic changes that cause cancer can happen because of: (1) errors that occur when cells divide; (2) damage to DNA that is caused by harmful environmental substances, such as certain pesticides or ultraviolet rays from the sun; or (3) inherited variants of genes that carry higher risk for cancer. Typically, our body has mechanisms to eliminate damaged genetic material before it becomes cancerous. Because this ability wanes as we age, and we are the culmination of additive environmental insults to our DNA, most cancers have a higher risk as we age.

There are many differences between cancerous cells and normal cells. For example, cancer cells continue to grow in the absence of the typical cellular signals promoting growth. Normal cells only grow when they receive such signals. Cancer cells ignore signals that tell cells to stop dividing or to die (a process called programmed cell death, or apoptosis), which allows them to invade adjacent tissue and spread to other areas of the body. By comparison, normal cells stop growing when they encounter other cells, and most normal tissue cells (other than blood) do not move around the body. Another very important feature of cancer cells is that they produce signals to promote the growth of blood vessels toward tumors. These new blood vessels supply oxygen and nutrients to, and remove waste products from, tumors, which in turn supports the tumors' growth. Cancer cells also often have the ability to hide from the immune system or even "trick" the immune system into helping cancer cells survive and multiply. The immune system typically works to eliminate damaged or abnormal cells. Unlike typical cells, cancer cells often accumulate changes in their chromosomes, including duplications and deletions of parts of the chromosomes. Indeed, some cancer cells have double the normal number of chromosomes, which means that more opportunities for mutations exist! Importantly, cell division requires a lot of energy. To support their unrelenting cell division, cancer cells typically rely on different types of nutrients and energy sources compared to normal cells, including increased uptake of glucose, enhanced rates of glutaminolysis (breaking down of glutamine), and synthesis of new fatty acids as well as anaerobic glycolysis. Indeed, cancer cells derive most of their energy from anaerobic glycolysis (a metabolic process that elevates blood glucose concentration and does not need oxygen to occur). Glycogen is a form of stored energy that is located in the liver and muscle tissues. Glycogen is typically pulled out of storage to be converted to glucose for fueling bodily functions prior to ingesting your first calories upon awakening. The blood glucose is converted to lactate for energy followed by lactate fermentation; this process occurs even when oxygen is not available. This lets cancer cells grow more quickly. Finally, cell division is firmly controlled by circadian rhythms in normal cells, especially cell cycle check points and cell cycle progression. Cell cycle check points are internal controls that monitor internal and external factors to "decide" that cell division can proceed. However, cancer cells often show temporal dysregulation that again allows continued growth and cell division across the day and ignores negative regulators during the check points.

The scientific community has considered disrupted circadian rhythms to be a potential risk factor for cancer since the late 20th century. Among other modern disruptors of circadian rhythms, such as social jet lag and night-shift work, the effects of artificial light exposure at night on cancer have garnered a lot of attention because of the increasing presence of light at night across the globe (Figure 6.1).

The endocrine system comprises chemical messengers called hormones that are secreted from specialized glands. There are

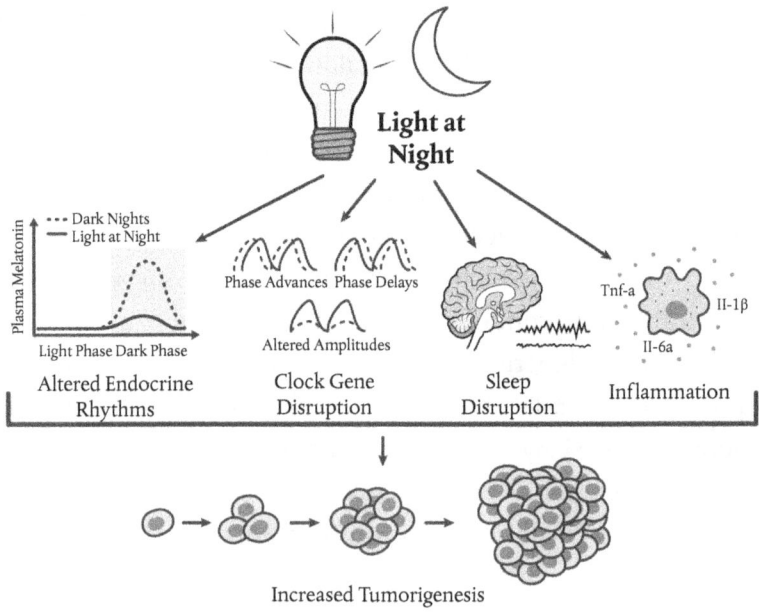

Figure 6.1 Summary of the effects of artificial light at night on cancer.

several types of hormones including steroid hormones. Estrogens are the class of steroids typically circulating at the highest blood concentrations in females, whereas androgens are the class of steroid hormones circulating in the blood of males. Several endocrine-related cancers have been described that represent a group of sex steroid responsive cancers, such as cancers of the breast, endometrium, prostate, and testis. The cells of estrogen receptor (ER)–positive breast cancer have receptors that allow them to use estrogens to promote their growth and division. The use of anti-estrogenic/anti-androgenic treatments for sex-hormone-responsive cancers has improved treatment options. The disruptive effects of artificial light at night on endocrine function have led to a primary focus on the association between exposure to artificial light at night and breast and prostate cancers. Again, because of the ethical issues of causative artificial light at night and cancer studies, our understanding of artificial light at night exposure as a risk factor for human cancer comes primarily from epidemiological observational, case-control, and cohort studies, as well as a robust set of experiments in preclinical studies with laboratory rodents.

As noted in previous chapters, night-shift workers often experience multiple and simultaneous forms of disrupted circadian rhythms, including mistimed food intake, social jet lag, sleep deprivation, and artificial light at night exposure. This makes it challenging to parse out the disruptive aspects of night-shift work that lead to harmful physiological consequences, such as increased cancer incidence. However, we will move through the various risk factors associated with disrupted circadian rhythms and cancer.

Jet Lag

Although a number of animal models of experimental jet lag have produced a link between disrupted circadian rhythms and cancer, only a few studies in humans have pointed to this link. Flight crews experience significant jet lag over their careers. In their work, they are exposed to a number of cancer risks including cosmic ionizing radiation, a form of radiation from space that intensifies at higher altitudes; this radiation exposure is higher for flight crews than for any other profession, and exposure levels are largely unregulated in the airline industry. Additionally, many flight attendants who are working today (in 2015) were exposed to significant secondhand smoke before it was illegal to smoke on flights. Importantly, flight crews are exposed to disrupted circadian rhythms due to night-shift work, irregular sleep schedules, and jet lag from crossing multiple time zones. On the other hand, the cabin crews tend to be younger adults and relatively physically fit, and the pilots undergo regular medical exams with aviation medical examiners. Overall, studies of cancer risk of flight crews have been uncommon and not particularly well designed. Considered together, however, these early studies suggest a correlation may exist between time on the job and in-flight experiences on the one hand and increased rates of breast and skin cancers, especially, and total cancer risk (the sum of cancers in any organ) on the other.[1-3]

In a relatively recent, well-designed study of >5,300 US flight attendants, participants were asked about their work habits and their health. This was part of the Harvard Flight Attendant Health Study that began in 2007.[4] The flight attendant data were compared to data from a similar group of people who were not flight crews and who were part of an annual survey conducted by the US Centers for Disease Control and Prevention. This comparison revealed that flight attendants had a higher incidence of every cancer type that was examined, especially breast cancer, melanoma skin cancer, and nonmelanoma skin cancer (basal cell carcinoma and squamous cell carcinoma) among female flight attendants. The other cancer types that were elevated included thyroid, cervical, endometrial, and gastrointestinal cancers. Specifically, 3.4% of female flight attendants reported having had breast cancer, compared with 2.3% of women in the general population group. Also, 7.4% of female flight attendants reported having had nonmelanoma skin cancer compared with 1.8% of women in the general population group! It is especially surprising that these cancer rates were elevated because of the low rates of smoking and obesity, common cancer risk factors, among flight crews compared to the general population.[4]

A study of pilots revealed a mild increase in relative risk for prostate cancer associated with number of flight hours.[5] The role of persistent jet lag on cancer in flight crews is consistent with the animal model data, which likewise show that jet lag promotes cancer growth.[6]

Social Jet Lag

Few studies of social jet lag on cancer incidence have been reported. One recent study, however, examined the role of social jet lag on prostate cancer risk in 7,455 cancer-free men in the Alberta's Tomorrow Project. In this study, participants were queried about the extent to which they experienced social jet lag and were followed for an average of 9.6 years; over this time, 250 men were diagnosed with prostate cancer. Importantly, men with the largest shifts in wake or sleep times caused by social jetlag experienced the highest risk for prostate cancer.[7]

Exposure to Dim Lighting During the Day

Few, if any studies in humans, have linked dim light during the day with cancer risks. One group of scientists conducted a study during which women undergoing chemotherapy treatment for breast cancer were exposed to either bright white light during the morning or dim red light. The results suggested that morning bright light treatment may prevent fatigue from worsening during chemotherapy, and may be a useful intervention during chemotherapy for breast cancer, but did not change fatigue compared to morning dim light exposure.[8] However, it is critically important for human entrainment of their circadian clock to be exposed to cycled lighting, ideally exposure to daytime sunlight and dark nights.

Late Night Food Intake

A study conducted in France reported a correlation between the risk of developing either breast or prostate cancer and the lateness of the final meal of the day.[9] Furthermore, the timing of food intake is important in cancer risks. For instance, the longer the interval between the last meal of the day and sleep onset, the lower the person's cancer risk.[10] Recall that cancer cells require more fuel to power their growth and division compared to normal cells. Diabetes is a risk factor for several cancers. Because of the elevated glucose levels in people with diabetes, the availability of this cellular fuel may partially explain this risk factor. As we learned in Chapter 2, late night food ingestion was associated with an elevated risk for diabetes. This risk factor might explain, in part, why night-shift workers are at increased risk for some types of cancer (see below).

Exposure to Light at Night

Exposure to artificial light at night has been associated with certain cancers, especially breast and prostate cancer. Multiple scientific panels of the World Health Organization (WHO) International Agency for Research on Cancer (IARC),[11] American Medical Association,[12] and NIH National Toxicology Program[13] have determined that exposure to light at night and insufficient light during the day increase the risk of breast cancer and other disorders.

Epidemiological data suggest that there is a slight, but positive, association between exposure to artificial light at night and cancer overall in humans. Most epidemiological studies examined breast cancer risk ratios, whereas several other studies reported on prostate and other cancers. Levels of environmental, artificial light at night are firmly associated with breast cancer incidence in humans. For example, a positive association between breast cancer and exposure to artificial light at night was observed in several satellite imaging studies. These studies examine the relationship between light pollution, typically caused by street and building lighting, and cancer. For instance, two separate studies of Israeli populations reported that urban artificial light at night was associated with increased risk for breast cancer.[14,15]

Another satellite study concluded that Georgian women exposed to "high" levels of artificial light at night had increased risk for developing breast cancer in comparison to women exposed to "low" levels.[16] Other US studies reported significant associations between levels of artificial light at night and breast cancer incidence.[17,18]

Two major global studies examining artificial light at night and breast cancer incidence have also been reported. Using the GLOBOCAN 2002 database, a significant association between environmental light at night and breast cancer was identified.[19] Another group conducted a follow-up study in which they stratified countries into Western, Gulf State, Southeast Asian, and "Other" categories; once again, an association between artificial light at night and the incidence of breast cancer was reported.[20] However, few persons are likely spending much time outdoors under street lighting at night, so other studies have attempted to understand the risk of more typical exposure to bedroom levels of artificial light at night on cancer risks.

One survey study reported that higher artificial light-at-night levels in one's bedroom were significantly associated with increased risk for breast cancer; unfortunately, assessment of bedroom lighting was relatively crude, asking participants to score their bedroom lighting levels from 0 ("completely dark") to 4 ("very strong light").[21] A study of women indicates that those who wake up and turn on lights during the night for more than two hours per night at least twice per week have an increased risk for breast cancer.[22] Although a few studies have reported no association between artificial light at night and breast cancer, the most comprehensive meta-analysis of 14 published studies concluded that both outdoor and indoor exposure to artificial light at night is associated with an increased risk for breast cancer.[23]

Most studies of light at night have focused on breast cancer, but other types of cancer have also been studied. For example, two analyses using satellite images of artificial light at night (one focused on outdoor exposure to short wavelength [blue] light) revealed an increased risk for prostate, but not lung, cancer in men.[24,25] Similarly, a positive correlation was reported between the amount of nighttime bedroom lighting levels and elevated risk for prostate cancer.[25] Although exposure to broad-spectrum artificial light at night is not associated with increased risk for colorectal cancer, exposure to short wavelength (425–560 nm— bluish) light *is* associated with elevated risk for colorectal cancer.

Considered together, the current evidence suggests a relatively strong association between artificial light at night and breast cancer. However, the relationship between artificial light at night and other cancers remains modest—much of this conclusion is based on satellite imaging data. But again, most people are not exposed to significant outdoor light at night, so satellite imaging studies have only modest value in pointing out hot spots for light at night and cancer risks. Future studies should be designed to actively measure nighttime light exposure for individuals in their homes or workplaces, possibly by using wearable light-sensors, to examine the relationship between disrupted circadian rhythms and other types of cancer.

Night-Shift Work

There is a large and robust scientific literature suggesting night-shift workers have increased risk for cancer compared to their day-shift counterparts. Indeed, night-shift work has been classified as a group 2a carcinogen (i.e., a probable carcinogen) for humans by the International Agency for Research on Cancer in 2007 and again in 2019.[26,27] Indeed, nurses working night shifts in some European countries receive "hazard pay" because of the health risks.[28]

Several epidemiological studies reveal that individuals involved in night-shift work are at a higher risk of developing cancer, including colorectal, rectal, endometrial, and breast cancers.[29-32] An extensive meta-analysis of over 30 separate studies revealed that a prolonged history of night-shift work is associated with increased risk of breast and prostate cancer. Another meta-analysis also revealed increased risk of breast cancer in premenopausal night-shift workers, especially those with extended exposure to night-shift work. Women who stopped working night shifts for two or more years reduced their odds of developing breast cancer.[33,34] However, some studies suggest that experience with night-shift work carries a persistent increase in breast cancer risk, even after night-shift working ends. For example, a study of more than 7,000 30- to 54-year-old Danish women indicated that individuals who had worked the night-shift for at least 18 months during their lives continued to display a higher risk for breast cancer.[35] A study of Polish women with breast cancer indicated that night-shift work was just below BMI as the most cancer-promoting factor. In this population, night-shift work increased breast cancer risk by 2.34 times.[36] Taking all the evidence together, prolonged night-shift work, especially if started at a young age, appears to be linked with a higher lifetime risk of breast cancer.[37,38]

Although the mechanisms underlying the effects of exposure to light at night and night-shift work on cancer risk have not been confirmed in humans, one hypothesis is that it is linked to melatonin suppression. Recall that exposure to light during the night inhibits melatonin expression, and laboratory studies in animals have shown that in addition to promoting sleep, melatonin displays robust suppression of tumors.[39,40] Thus, circadian disruption may increase cancer risk by diminishing natural antitumor defenses.

There have literally been hundreds of studies in rodents showing that disruption of circadian rhythms, either by exposure to light at night, simulated jet lag, or night-shift studies, increases the risk of developing cancer, enhances the odds of developing tumors, and promotes metastases of tumors once they develop. Thus, in contrast to clinical studies relying on correlations, compelling animal model data indicate that disrupted circadian rhythms have causative effects on cancer.

Many of the effects of light at night on night-shift workers have been attributed to sleep disruption. One important advantage to examining nocturnal mice and rats is that they sleep during the day; light at night disrupts circadian rhythms without typically affecting sleep in these animals. Thus, the effects of sleep and disrupted circadian rhythms can be separated in these studies. To summarize these preclinical studies, in the absence of disrupted sleep, light at night increases cancer risks.

The first example that demonstrated an association between light at night and cancer was published in 1964.[41] This study described a strain of mice that develop spontaneous mammary tumors; female mice of this strain displayed more tumors when exposed to around-the-clock light.[41] A related strain of mice

displayed elevated numbers of liver carcinoma, lung adeno-
carcinoma, and leukemias as well as shortened lifespans when
exposed to chronic light.[42] Other studies reported that laboratory rats that were exposed to both continuous light from
birth and a chemical carcinogen displayed increased mammary
tumors relative to rats housed in typical light-dark cycles.[43] Treatment of these rats with melatonin in a dosing regimen to mimic
typical nocturnal secretion completely reversed the elevated
rate of tumorigenesis! Housing male and female rats in constant
light conditions from one month of age caused them to display
elevated spontaneous tumorigenesis, accelerated aging, and
reduced rates of survival.[44] However, if the constant lighting conditions did not start until the rats were 14 months of age, there
was only evidence of increased tumors in female rats and the
lighting conditions did not affect survival in either sex. In other
words, exposure to constant lights evokes a favorable environment for tumor development. For example, exposure to constant
light for just five weeks elevated blood glucose concentrations
and increased glioma tumor growth in rats. As mentioned earlier, increased glucose availability promotes tumor growth.
Constant light conditions also influence the tumor microenvi-
ronment to promote inflammation, glucose update molecules,
and development of blood vessel formation to and within the
tumors.[44] Considered together, the evidence from preclinical
studies indicates that exposure to constant light throughout the
night has a dramatic effect on tumor development, growth, and
host mortality.

Constant light throughout the day and night not only disrupts
circadian rhythms but actually masks their influence on physiology and behavior. In response to the clinical studies indicating

that disrupted circadian rhythms promote cancer, recent pre-clinical studies have focused on providing low levels of light at night. A more ecologically grounded experiment studied rats housed in typical light-dark cycles or exposed to 12 hours of light (300 lux) and 12 hours of dim (2x full moon light, 0.21 lux). After 12 weeks, rats exposed to dim light at night displayed reduced markers of melatonin, as well as reduced survival rates compared to rats housed under dark nights.[45]

The role of melatonin as an antitumor hormone has received much support in preclinical studies. Recall that melatonin is secreted only at night, both in diurnal (humans) and nocturnal (mice and rats) creatures (see Chapter 2, Box 2.1). In a series of remarkable studies, the role of artificial dim light at night on melatonin and tumor development and growth was examined. In one study, rats were exposed to dim light at night (0.2 lux) or dark nights for 2–3 weeks after implantation of the same number of human steroid-receptor-negative tumor cells. The rats exposed to dim artificial light at night displayed accelerated growth of these human derived cancer cells.[45] Rats exposed to typical cycling light-dark patterns expressed maximal markers of cell prolif-eration (i.e., tumor growth) during the early light phase, but tended to be relatively low throughout the day. Consistent with accelerated cell proliferation, rats housed under dim artificial light at night conditions continuously expressed high levels of cell proliferation molecular markers.[45] Other molecular signals were also disrupted by exposure to light at night in this study. Considered together, these results demonstrate that exposure to artificial dim light at night accelerates tumor growth, in part through continuous activation of a specific molecular pathway associated with cell proliferation.

In another clever experiment, this same group examined the effects of artificial dim light at night and human plasma on rats implanted with a xenograft of human breast tumor cells or a rat liver cancer tumor.[46] (A xenograft refers to a tissue or organ that is derived from a species that is different from the recipient of the specimen.) Blood samples were obtained from premenopausal women either during the daytime (melatonin-deficient), nighttime (melatonin-infused), or nighttime after 90 min exposure to bright lights (melatonin-deficient). Xenografts perfused with blood containing no melatonin (either daytime or exposed to light at night) displayed increased tumor growth of both the liver tumors and steroid receptor-negative breast tumors compared to xenografts perfused with melatonin-rich blood.

This research team also reported that tissue-isolated rat liver cancer was highly responsive to melatonin; treatment with melatonin at physiological concentrations decreased tumor proliferative activity.[47,48] Similarly, suppression of melatonin concentrations in rats by exposure to artificial light at night promotes metastases of implanted human breast cancer tumors into liver, lung, and brain.[49]

So, what are the molecular mechanisms underlying how dys-regulated circadian rhythms influence cancer? The answer to that question remains largely unknown. However, numerous tantalizing hints have been revealed by examining the bidirectional role of core circadian clock genes and the cell cycle in both animal models and cancer patients. For example, many breast cancer patients display mutations and increased methylation of the circadian clock gene promoters in Per1 and Per2.[50,51] Expression of two proteins critical for circadian clock function (PERs and CRYs) has been reported to be decreased within breast tumors relative to surrounding noncancerous breast tissue.[52,53] Similarly, reduced expression of one or more PERs has been reported in patient samples of colorectal, prostate, adrenal, ovarian, endometrial, lung, glioma, and pancreatic cancer.[54]

Together, the results of these preclinical studies demonstrate that melatonin may be anti-oncogenic and that suppression of melatonin by light exposure at night is sufficient to change the tumor cell growth parameters in rodents bearing human tumors. Recent research has continued to explore the effects of light at night on cancer by focusing on the spectral composition (wavelength) of this light. Short wavelength (blue) light increased tumor growth, promoted lung metastases, and accelerated DNA hypomethylation (which affects cell division mediators and can lead to cancer) in mice. Melatonin treatment of mice exposed to light at night ameliorated the effects of nighttime lighting on tumor development and growth.[55]

In addition to the bad news about accelerating tumor development and metastases, exposure to dim light at night can also have negative effects on cancer treatment by chemotherapies. Several studies have reported that exposure to artificial light at night promotes resistance to common chemotherapeutic drugs including doxorubicin, paclitaxel, and the hormonal therapy tamoxifen.[56,57] In preclinical studies, exposure to artificial light at night increases carcinogenesis, but the mechanisms remain elusive. Whether light at night promotes tumors by altering clock gene expression, reducing natural melatonin secretion, or—what is likely—combining both mechanisms, remains to be determined. However, it is apparent that there are beneficial effects of melatonin on tumor initiation and progression. Translating these preclinical results to humans is a high priority.

Summary

Thus, an evidence-based strategy to reduce cancer risks includes maintaining a healthy body weight and synchronizing one's circadian rhythms to the solar day. Avoid light at night, but seek out bright light during the day. Avoid late night food intake. Fast between dinner and breakfast, preferably 14 hours. The same especially holds true for night-shift workers. As I told my friend Jennifer, the night-shift nurse, it is important for night-shift workers to be aware of the potential negative impacts on their health and make an effort to maintain a healthy circadian organization and diet. This may involve completely flipping their days and nights. When they arrive at home after the night shift, they

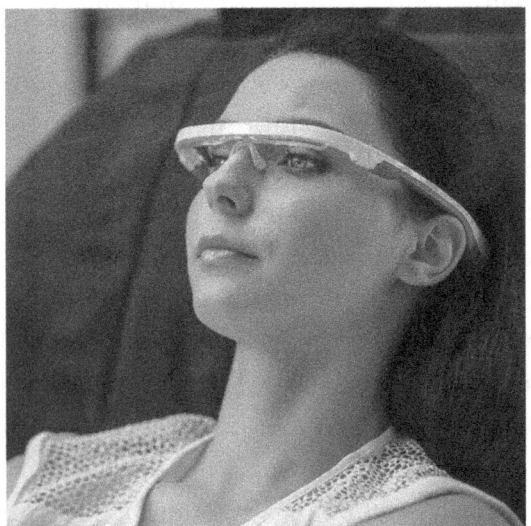

Figure 6.2 Blue light therapy visors can be used early during night shifts to help set (entrain) circadian rhythms. Image used with permission from Ayo.

should spend some time in low lighting before going to bed. Use black-out curtains or sleep masks to make the days similar to nights. Try to be exposed to bright light while at work. Special blue-enriched light visors can help with this (Figure 6.2). Night-shift workers should try to block out short wavelength (blue) light with special blue-light-blocking goggles on their way home from work. And in consultation with their physician, they can increase their natural melatonin concentrations with melatonin supplements taken at night.

References

1. Tokumaru O, Haruki K, Bacal K, Katagiri T, Yamamoto T, Sakurai Y. 2006. Incidence of cancer among female flight attendants: a meta-analysis. *Journal of Travel Medicine*, 13: 127–132.

2. Reynolds P, Cone J, Layefsky M, Goldberg DE, Hurley S. 2002. Cancer incidence in California flight attendants (United States). *Cancer Causes Control*, 13: 317–324.

3. Pukkala E, Aspholm R, Auvinen A, Eliasch H, Gundestrup M, Haldorsen T, et al. 2002. Incidence of cancer among Nordic airline pilots over five decades: occupational cohort study. *British Medical Journal*, 325: 567.

4. McNeely, E., Mordukhovich, I., Staffa, S, Tideman S, Gale S, Coull B. 2018. Cancer prevalence among flight attendants compared to the general population. *Environmental Health*, 17: 49.

5. Pukkala E, Aspholm R, Auvinen A, Eliasch H, Gundestrup M, Haldorsen T, Hammar N, Hrafnkelsson J, Kyyrönen P, Linnersjö A, Rafnsson V, Storm H, Tveten U. 2003. Cancer incidence among 10,211 airline pilots: a Nordic study. *Aviation, Space and Environmental Medicine*, 74: 699–706.

6. Filipski E, Delaunay F, King VM, Wu MW, Claustrat B, Gréchez-Cassiau A, Guettier C, Hastings MH, Francis L. 2004. Effects of chronic jet lag on tumor progression in mice. *Cancer Research*, 64: 7879–7885.

7. Hu L, Harper A, Heer E, McNeil J, Cao C, Park Y, Martell K, Gotto G, Shen-Tu G, Peters C, Brenner D, Yang L. 2020. Social jetlag and prostate cancer incidence in Alberta's Tomorrow Project: a prospective cohort study. *Cancers*, 12: 3873.

8. Ancoli-Israel S, Rissling M, Neikrug A, Trofimenko V, Natarajan L, Parker BA, Lawton S, Desan P, Liu L. 2012. Light treatment prevents fatigue in women undergoing chemotherapy for breast cancer. *Supportive Care in Cancer*, 20: 1211–1219.

9. Pham TT, Lee ES, Kong SY, Kim J, Kim SY, Joo J, Yoon KA, Park B. 2019. Night-shift work, circadian and melatonin pathway related genes and their interaction on breast cancer risk: evidence from a case-control study in Korean women. *Scientific Reports*, 9: 10982.

10. Kogevinas M, Espinosa A, Castelló A, et al. 2018. Effect of mistimed eating patterns on breast and prostate cancer risk (MCC-Spain Study). *International Journal of Cancer*, 143: 2380–2389.

11. International Agency for Research on Cancer. 2010. Working group on the evaluation of carcinogenic risks to humans: shift work. *Painting, Firefighting, and Shiftwork*, 98: 563–764.

12. Stevens RG, Brainard GC, Blask DE, Lockley SW, Motta ME. 2013. Adverse health effects of nighttime lighting: comments on American Medical Association policy statement. *American Journal of Preventative Medicine*, 45: 343–346.

13. US National Toxicology Program. NTP 2021. Cancer hazard assessment report on night shift work and light at night. National Toxicology Program, US Department of Health and Human Services. https://ntp.niehs.nih.gov/ntp/results/pubs/cancer_assessment/lanfinal20210400_508.pdf.

14. Kloog I, Haim A, Portnov BA. 2009. Using kernel density function as an urban analysis tool: investigating the association between nightlight exposure and the incidence of breast cancer in Haifa, Israel. *Computers, Environment and Urban Systems*, 33: 55–63.

15. Kloog I, Haim A, Stevens RG, Barchana M, Portnov BA. 2008. Light at night co-distributes with incident of breast, but not lung, cancer in the female population of Israel. *Chronobiology International*, 25: 65–81.

16. Bauer SE, Wagner SE, Burch J, Bayakly R, Vena JE. 2013. A case-referent study: light at night and breast cancer risk in Georgia. *International Journal of Health Geography*, 12: 1–10.

17. James P, Bertrand KA, Hart JE, Schernhammer ES, Tamimi RM, Laden F. 2017. Outdoor light at night and breast cancer incidence in the Nurses' Health Study II. *Environmental Health Perspectives*, 125: 87010.

18. Hurley S, Goldberg D, Nelson D, Hertz A, Horn-Ross PL, Bernstein L, Reynolds P. 2014. Light at night and breast cancer risk among California teachers. *Epidemiology*, 25: 697–706.

19. Kloog I, Stevens RG, Haim A, Portnov BA. 2010. Nighttime light level co-distributes with breast cancer incidence worldwide. *Cancer Causes Control*, 21: 2059–2068.

20. Rybnikova N, Haim A, Portnov BA. 2015. Artificial light at night (ALAN) and breast cancer incidence worldwide: a revisit of earlier findings with analysis of current trends. *Chronobiology International*, 32: 757–773.

21. Kloog I, Portnov BA, Rennert HS, Haim A. 2011. Does the modern urbanized sleeping habitat pose a breast cancer risk? *Chronobiology International*, 28: 76–80.

22. O'Leary ES, Schoenfeld ER, Stevens RG, Kabat GC, Henderson K, Grimson R, Gammon MD, Leske MC. 2006. Shift work, light at night, and breast cancer on Long Island, New York. *American Journal of Epidemiology*, 164: 358–366.

23. Lai KY, Sarkar C, Ni MY, Cheung LWT, Gallacher J, Webster C. 2020. Exposure to light at night (LAN) and risk of breast cancer: a systematic review and meta-analysis. *Science of the Total Environment*, 143159.

24. Kloog I, Haim A, Stevens RG, Portnov BA. 2009. Global co-distribution of light at night (LAN) and cancers of prostate, colon, and lung in men. *Chronobiology International*, 26: 108–125.

25. Garcia-Saenz A, Sánchez de Miguel A, Espinosa A, et al. 2018. Evaluating the association between artificial light-at-night exposure and breast and prostate cancer risk in Spain (MCC-Spain Study). *Environmental Health Perspectives*, 126: 47011.

26. Straif K, et al. 2006. Carcinogenicity of shift-work, painting, and firefighting. *Lancet Oncology*, 8: 1065–1066.

27. Ward EM, et al. 2019. Carcinogenicity of night shift work. *Lancet Oncology*, 20: 1058–1059.

28. Gärtner J, Rosa RR, Roach G, Kubo T, Takahashi M. 2019. Working Time Society consensus statements: regulatory approaches to reduce risks associated with shift work-a global comparison. *Indian Health*, 57: 245–263.

29. Schernhammer ES, et al. 2001. Rotating night shifts and risk of breast cancer in women participating in the Nurses' Health Study. *Journal of the National Cancer Institute*, 93: 1563–1568.

30. Viswanathan AN, Hankinson SE, Schernhammer ES. 2007. Night shift work and the risk of endometrial cancer. *Cancer Research*, 67: 10618–10622.

31. Schernhammer ES, et al. 2003. Night-shift work and risk of colorectal cancer in the nurses' health study. *Journal of the National Cancer Institute*, 95: 825–828.

32. Wegrzyn LR, et al. 2017. Rotating night-shift work and the risk of breast cancer in the nurses' health studies. *American Journal of Epidemiology*, 186: 532–540.

33. Cordina-Duverger E, et al. 2018. Night shift work and breast cancer: a pooled analysis of population-based case–control studies with complete work history. *European Journal of Epidemiology*, 33: 369–379.

34. Schernhammer ES, Kroenke CH, Laden F, Hankinson SE. 2006. Night work and risk of breast cancer. *Epidemiology*, 17: 108–111.

35. Hansen, J. 2001. Increased breast cancer risk among women who work predominantly at night. *Epidemiology*, 12: 74–77.

36. Szkiela M, Kusideł E, Makowiec-Dąbrowska T, Kaleta D. 2020. Night shift work: a risk factor for breast cancer. *International Journal of Environment Research and Public Health*, 17: 6594.

37. Wegrzyn LR, Tamimi RM, Rosner BA, Brown SB, Stevens RG, Eliassen AH, Laden F, Willett WC, Hankinson SE, Schernhammer ES. 2017. Rotating night-shift work and the risk of breast cancer in the nurses' health studies. *American Journal of Epidemiology*, 186: 532–540.

38. Gómez-Salgado J, Fagundo-Rivera J, Ortega-Moreno M, Allande-Cussó R, Ayuso-Murillo D, Ruiz-Frutos C. 2021. Night work and breast cancer risk in nurses: multifactorial risk analysis. *Cancers*, 13: 1470.

39. Reiter RJ, Rosales-Corral SA, Tan DX, Acuna-Castroviejo D, Qin L, Yang SF, Xu K. 2017. Melatonin, a full service anti-cancer agent: inhibition of initiation, progression and metastasis. *International Journal of Molecular Sciences*, 18: 843.

40. Lu KH, Su SC, Lin CW, Hsieh YH, Lin YC, Chien MH, Reiter RJ, Yang SF. 2018. Melatonin attenuates osteosarcoma cell invasion by suppression of C-C motif chemokine ligand 24 through inhibition of the c-Jun N-terminal kinase pathway. *Journal of Pineal Research*, 65: e12507.

41. Jöchle, W. 1964. Trends in photophysiologic concepts. *Annals of the New York Academy of Science*, 117: 88–104.

42. Anisimov VN, Baturin DA, Popovich IG, Zabezhinski MA, Manton KG, Semenchenko AV, Yashin AI. 2004. Effect of exposure to light-at-night on life span and spontaneous carcinogenesis in female CBA mice. *International Journal of Cancer*, 111: 475–479.

43. Mhatre MC, Shah PN, Juneja HS. 1984. Effect of varying photoperiods on mammary morphology, DNA synthesis, and hormone profile in female rats. *Journal of the National Cancer Institute*, 72: 1411–1416.

44. Vinogradova IA, Anisimov VN, Bukalev AV, Ilyukha VA, Khizhkin EA, Lotosh TA, Semenchenko AV, Zabezhinski MA. 2010. Circadian disruption induced by light-at-night accelerates aging and promotes tumorigenesis in young but not in old rats. *Aging*, 2: 82–92.

45. Cos S, Mediavilla D, Martínez-Campa C, González A, Alonso-González C, Sánchez-Barceló EJ. 2006. Exposure to light-at-night increases the growth of DMBA-induced mammary adenocarcinomas in rats. *Cancer Letters*, 235: 266–271.

46. Blask DE, Brainard GC, Dauchy RT, Hanifin JP, Davidson LK, Krause JA, Sauer LA, Rivera-Bermudez MA, Dubocovich ML, Jasser SA, Lynch DT, Rollag MD, Zalatan F. 2005. Melatonin-depleted blood from premenopausal women exposed to light at night stimulates growth of human breast cancer xenografts in nude rats. *Cancer Research*, 65: 11174–11184.

47. Blask DE, Sauer LA, Dauchy RT, Holowachuk EW, Ruhoff MS, Kopff HS. 1999. Melatonin inhibition of cancer growth in vivo involves suppression of tumor fatty acid metabolism via melatonin-receptor-mediated signal transduction events. *Cancer Research*, 59: 463–470.

48. Audinot V, Mailliet F, Lahaye-Brasseur C, Bonnaud A, Le Gall A, Amossé C, Dromaint S, Rodriguez M, Nagel N, Galizzi JP, Malpaux B, Guillaumet G, Lesieur D, Lefoulon F, Renard P, Delagrange P, Boutin JA. 2003. New selective ligands of human cloned melatonin MT1 and MT2 receptors. *Naunyn Schmiedebergs Archives in Pharmacology*, 367: 553–561.

49. Mao L, Summers W, Xiang S, Yuan L, Dauchy RT, Reynolds A, Wren-Dail MA, Pointer D, Frasch T, Blask DE, Hill SM. 2016. Melatonin represses metastasis in Her2-positive human breast cancer cells by suppressing RSK2 expression. *Molecular Cancer Research*, 14: 1159–1169.

50. Gery S, Koeffler HP. 2010. Circadian rhythms and cancer. *Cell Cycle*, 9: 1097–1103.

51. Sjöblom T, Jones S, Wood LD, et al. 2006. The consensus coding sequences of human breast and colorectal cancers. *Science*, 314: 268–274.

52. Chen ST, Choo KB, Hou MF, Yeh KT, Kuo SJ, Chang JG. 2005. Deregulated expression of the PER1, PER2 and PER3 genes in breast cancers. *Carcinogenesis*, 26: 1241–1246.

53. Homan AE, Zheng T, Yi CH, Stevens RG, Ba Y, Zhang Y, Leaderer D, Holford T, Hansen J, Zhu Y. 2010. The core circadian gene cryptochrome 2 influences breast cancer risk, possibly by mediating hormone signaling. *Cancer Prevention Research*, 3: 539–548.

54. Walker WH, Bumgarner JR, Walton JC, Liu JA, Melendez-Fernandez OH, Nelson RJ, DeVries AC. 2020. Light pollution and cancer. *International Journal of Molecular Sciences*, 21: 9360.

55. Zubidat AE, Fares B, Fares F, Haim A. 2018. Artificial light at night of different spectral compositions differentially affects tumor growth in mice: interaction with melatonin and epigenetic pathways. *Cancer Control*, 25: 1073274818812908.

56. Dauchy RT, Xiang S, Mao L, Brimer S, Wren MA, Yuan L, Anbalagan M, Hauch A, Frasch T, Rowan BG, Blask DE, Hill SM. 2014. Circadian and melatonin disruption by exposure to light at night drives intrinsic resistance to tamoxifen therapy in breast cancer. *Cancer Research*, 74: 4099–4110.

57. Xiang S, Dauchy RT, Hauch A, Mao L, Yuan L, Wren MA, Belancio VP, Mondal D, Frasch T, Blask DE, Hill SM. 2015. Doxorubicin resistance in breast cancer is driven by light at night-induced disruption of the circadian melatonin signal. *Journal of Pineal Research*, 59: 60–69.

7

LIGHT AND CARDIAC FUNCTION

When thinking of rhythmic bodily functions, the thumping of our hearts comes to mind first for many of us. Not only is there the steady rhythm of our pulse but also our pulse rate shows a general time-of-day pattern that is driven by our circadian clocks. Typically, pulse rates are low during the night when there is limited physical activity, steadily increases right before awakening, and then continues to increase as we become more active. This predictable increase is reflective of both a programmed increase in heart rate that occurs in anticipation of the start of the active part of the day and in response to the onset of locomotor activity. The programmed increase in heart activity is a two-edged sword. It gets blood pumping to our muscles as our bodies prepare to forage for food—to break the nightly fast (breakfast). But if our coronary arteries, the blood vessels that supply the heart with the oxygen needed to power it, are clogged by cholesterol plaques, then the increased pulse rates can trigger heart problems. This chapter will explore how

Dark Matters. Randy J. Nelson, Oxford University Press. © Oxford University Press (2025).
DOI: 10.1093/9780197639979.003.0007

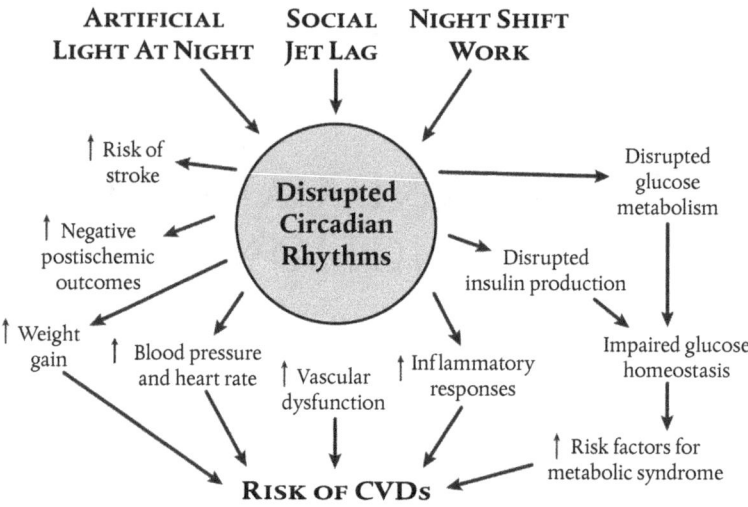

Figure 7.1 Bidirectional effects of disruption of circadian rhythms and risk factors for cardiovascular diseases (CVDs).

circadian rhythms and disrupted circadian rhythms influence cardiovascular function and disease (Figure 7.1).

Our nervous system is broadly organized into the central nervous system (CNS) and the peripheral nervous system. The peripheral nervous system can be further subdivided into the somatic nervous system (sometimes called the voluntary nervous system, which regulates muscular function) and the autonomic nervous system (regulating all the involuntary processes of our bodies). When you voluntarily raise your left arm to wave to a neighbor, you are using your somatic nervous system. When your adrenal glands produce more adrenaline to increase your heart rate and number of breaths when you see a car veer into your lane, that is the work of your autonomic nervous system. The autonomic nervous system comprises two components as well: the sympathetic and the parasympathetic branches. The sympathetic branch is activated in the "fight or flight" response. This system is typically activated when you perceive danger— your pulse rate, blood pressure, and breathing rate all increase to ready your muscles to escape or fight off the danger. Blood is diverted from reproductive organs, digestion, and immune tissues to muscles. You can breed, digest your meal, and build up immunity later . . . if there is a later! This is a classic stress response. Of course, we cannot remain in fight or flight mode throughout the day—it is too energetically demanding, and if the stress response continues, then cardiovascular, digestive, reproductive, and immune dysfunction can result. The parasympathetic branch of the autonomic nervous system regulates the so-called rest and digest responses. After the danger passes, then the parasympathetic branch reduces cardiac responses, blood pressure, and breathing rates, and blood flow is returned to the digestive tract, reproductive processes, and immune organs. People who meditate can activate their parasympathetic systems to promote relaxation. The heart and blood vessels are regulated by sympathetic and parasympathetic systems. When we wake up in the morning, our sympathetic systems are activated as we begin to forage for food; mid-day and later in the evening, our parasympathetic systems predominate.

Circadian regulation of vascular and endothelial (cells lining blood vessels) function is well documented.[1] Appropriate daily alignment of vascular function relies on a precisely regulated temporal balance between the parasympathetic and sympathetic branches of the autonomic nervous system to modulate heart rate, vascular tone, blood pressure, and endothelial function.[2] This temporal organization permits production of vascular mediators such as prothrombotic (clotting) and antithrombotic (nonclotting) factors, and nitric oxide (NO) (vasodilator) at the optimal time of day, generally at the start of the day for humans. Indeed, for chest pain (angina pectoralis) some folks are prescribed nitroglycerin, which is an artificial vasodilator that works by providing NO when needed to dilate coronary blood vessels. Daily anticipation of the increased cardiovascular demands (driven primarily by the sympathetic arm of the autonomic nervous system) of individuals' active periods, and decreased cardiovascular demands (mediated by the parasympathetic arm) during the inactive period optimizes overall biological function. Dysregulation of this balance between sympathetic and parasympathetic tone leads to cardiovascular and cerebrovascular diseases. It is difficult to fall asleep at night if our sympathetic nervous system is increasing our pulse and breathing rates. Of course, worrying about a test or work assignment due the next day can override the natural circadian variation in the activation and resting of our cardiovascular systems.

Each morning, sympathetic input drives increased blood pressure prior to awakening in people.[3,4] Blood pressure starts to increase throughout the morning; in humans, elevated blood pressure typically occurs between about 6:30 and 8:00 a.m. This time of day typically coincides with the increased expression of blood clotting factors, platelet production, and increased risk for thrombogenesis (formation of a blood clot). Blood pressure begins a gradual decrease in the afternoon, reaching a trough during the night. Failure to decrease blood pressure during the night is likely a result of the lack of vagal tone regulation by the sympathetic and parasympathetic systems. Vagal tone represents the activity of the vagus nerve and serves as a fundamental component of the parasympathetic branch. The vagus nerve receives both parasympathetic and sympathetic inputs, but its baseline output is highly parasympathetic and drives "rest and digest" responses. If the vagal tone is not properly regulated, then cardiac and vascular function does not "rest." Individuals who do not decrease their blood pressure at night are referred to as "nondippers" because their blood pressure remains relatively constant (and high) throughout the day and night. Nondipper status, in conjunction with the rhythm of expression of the clotting factors, places individuals at increased risk of pathological cardiac and cerebrovascular events during the morning.[2] Indeed, heart attacks and strokes that occur during the morning tend to be more severe and are associated with elevated diastolic blood pressure, increased hospital stay, and worse outcomes, including mortality, compared to strokes that occur later in the day. Furthermore, disrupted circadian rhythms are linked to higher risk for cardiac problems and strokes, and disrupted circadian rhythms are reported to play a role in stroke outcomes.[2] For example, night-shift work changes this precisely orchestrated and programmed increase in blood pressure, clotting factors, platelet production, and thrombogenesis.

Although briefly exceeded by COVID-19 deaths in early 2021, according to the US Centers for Disease Control (CDC), cardiovascular diseases continue to be the leading cause of mortality in the United States.[5] The most recent reports by the World Health Organization (WHO) attribute nearly a quarter of worldwide deaths to ischemic heart disease and stroke.[6] Ischemia is a condition in which blood flow and thus oxygen is restricted or reduced to a part of the body. Cardiac ischemia is decreased blood flow and oxygen to the heart muscles, whereas an ischemic stroke occurs in response to decreased blood and oxygen traveling to the brain. Cardiovascular and cerebrovascular diseases share multiple risk factors, many of which are modifiable, such as poor diet, obesity, high sodium intake, high blood pressure, and low physical activity. Another modifiable risk factor that has recently been identified is exposure to artificial light at night. As you now appreciate, artificial light at night disrupts circadian rhythms and interferes with the temporal organization of physiology and behavior, including factors associated with heart and brain blood vessel health.

One of the key mechanisms linking disrupted circadian rhythms and cardiac function is the autonomic nervous system, which controls the balance between the sympathetic (fight-or-flight) and parasympathetic (rest-and-digest) branches. The autonomic nervous system is a subsystem of the peripheral nervous system, which comprises the neural tissue in your body excluding your brain and spinal cord (which are part of your central nervous system). During the day, the sympathetic system is more dominant, promoting higher heart rate and blood pressure to accommodate increased activity, whereas the parasympathetic system dominates during sleep, leading to decreased heart rate and blood pressure. Disrupted circadian rhythms can alter this balance between sympathetic and parasympathetic activity, leading to an imbalance in autonomic regulation of the heart and vasculature, which can then lead to increased risk of negative cardiovascular and cerebrovascular events.

Several studies suggest that individuals with disrupted circadian rhythms, such as night-shift workers or those with sleep disorders, are more prone to developing cardiovascular and cerebrovascular conditions, including hypertension (high blood pressure), arrhythmias (irregular heart rhythms), and even heart disease. Arterial plaque can build up in coronary and brain arteries. A plaque forms when cholesterol builds up in the inner lining of the artery, and this process can develop into a condition called atherosclerosis. Atherosclerosis reduces the flow of blood through vessels and can, in turn, develop into coronary artery disease. Disrupted sleep patterns and irregular sleep-wake cycles can contribute to these adverse cardiac effects. Additionally, disruptions to circadian rhythms can affect the production and release of hormones involved in cardiovascular regulation. For example, melatonin, a hormone that helps regulate sleep-wake cycles, has been reported to have beneficial effects on cardio-vascular and cerebrovascular health. Disruptions in melatonin production, such as those experienced by individuals working night shifts or exposed to excessive artificial light during the night, can impair its protective effects on the vasculature.

In this chapter, I will briefly review the evidence that supports the hypothesis that circadian rhythm influences the cardiovascu-lar system and that disrupted circadian rhythms impair cardio-vascular function leading to more myocardial infarctions (heart attacks) and strokes.

Jet Lag

Although jet lag disrupts circadian rhythms, as we have seen, there have been limited reports linking jet lag among travelers or flight crews who cross multiple time zones with negative car-diovascular or cerebrovascular events. This may reflect that flight crews, who experience the most frequent jet lag, are typically in good physical condition. But there are few reports of a spike in heart attacks in passengers who have recently traveled multiple time zones in the scientific literature, and those few cases are attributed to disrupted sleep rather than disrupted circadian rhythms per se.

As mentioned, circadian rhythms play a crucial role in regulating physiological processes, including cardiovascular and cerebrovascular function. These rhythms help to coordinate the timing of heart rate, blood pressure, and other cardiovascular parameters throughout the day.[2] When jet lag occurs, the sudden shift in time zones can disrupt this coordinated regulation, leading to several cardiovascular changes.

Jet lag often disturbs sleep, including causing difficulty falling asleep or staying asleep during the night and excessive daytime sleepiness. Sleep disturbances have been associated with an increased risk of cardiovascular problems, including hypertension and heart disease. Disrupted sleep can also affect the functioning of the autonomic nervous system, which regulates cardiovascular and cerebrovascular activities, further contributing to unwelcome cardiovascular and cerebrovascular changes.[7]

Recall that melatonin is a hormone that typically is secreted at night and helps to regulate the sleep-wake cycle. Melatonin has also been linked to good cardiovascular health. When traveling across multiple time zones, the production and release of melatonin may be disrupted, leading to altered blood pressure regulation and endothelial function. This disruption could contribute to an increased risk of negative cardiovascular and cerebrovascular events among at-risk individuals.[7]

However, air travel itself, and not necessarily jet lag, increases the risk of developing blood clots in the veins of the legs, which can then enter the bloodstream and block an artery in the lungs, a condition called pulmonary embolism. In some very rare cases, the blood clot enters the arteries of the brain, causing a stroke. In one study that took place over an eight-year period, French researchers examined passengers who had a pulmonary embolism when they arrived at Charles de Gaulle airport in Paris. These patients were transported to a hospital by ambulance. Of the 155 million passengers who traveled through the airport during that period, only 65 people displayed a pulmonary embolus. Of those 65 individuals, 4 people (6%) suffered subsequent strokes, emphasizing the rarity of this condition.[8]

To manage the effects of jet lag on cardiovascular function, it is recommended to:

1. Stay hydrated: Drink plenty of water to maintain proper hydration during travel.
2. Gradually adjust to the new time zone: Gradually shift your sleep and activity schedule closer to the destination time zone before and during travel.
3. Optimize sleep hygiene: Create a sleep-friendly environment, maintain a regular sleep schedule, and consider using techniques such as exposure to bright light during the day or melatonin supplements during the early night consistent with the time at your destination to help regulate sleep patterns.[9]
4. Stay active: Engage in light physical activity or exercise upon arriving at the destination to help regulate circadian rhythms and promote overall cardiovascular health.[10]

5. Seek medical advice if necessary. If you have preexisting cardiovascular conditions or concerns, then consult with your physician before traveling to discuss any specific precautions or recommendations.

It is worth mentioning that this information is based on a general understanding of the impact of jet lag on cardiovascular function. For more detailed and personalized information, it is always best to consult with a healthcare professional who can provide specific guidance based on your medical history and circumstances.

Social Jet Lag

Social jet lag has been reported to affect metabolic disorders, sleep, obesity, and type 2 diabetes,[11] all of which contribute to cardiovascular disorders. Heart rate variability is often used as a proxy for the balance between sympathetic and parasympathetic function. Heart rate variability is a precise measure of the variation in time between each heartbeat. This variation is controlled by the autonomic nervous system. The changes between heartbeats are usually too small to detect without specialized monitoring devices. Generally, high heart rate variability is a sign of low stress (high parasympathetic activation or tone), whereas a low heart rate variability rate indicates high stress (high sympathetic activation or tone). In one study, high heart rate variability was assessed among individuals with low social jet lag shifts and high social jet lag shifts. During the first three hours of sleep the group with high social jet lag had lower high-frequency heart rate variability on a work day, as compared to a free day, whereas the high heart rate variability values of the low social jet lag group remained consistent between the work and free day.[12] Although it may seem counterintuitive, high heart rate variability is considered a sign of good vascular health.

Another related study of 145 healthy participants reported that people who experienced more than 2 hours of social jet lag displayed elevated concentrations of cortisol (the so-called stress hormone), slept less, and were more physically inactive than study participants who experienced less than 1 hour of social jet lag. Furthermore, study participants with more than two hours of social jet lag displayed increased resting pulse rates compared to those individuals with less than one hour of social jet lag.[13]

A study of Portuguese blue-collar workers emphasized the dangers of social jet lag on cardiovascular health. In addition to sociodemographic, health, and work productivity data, several parameters associated with cardiac health, such as blood pressure and cholesterol levels, were measured in these workers. After estimating cardiovascular risk with the Systematic Coronary Risk Evaluation (SCORE) chart, comparisons were made among individuals with different amounts of social jet lag.[14] Cardiovascular risk was high in 20% of the participants, but additional analyses revealed that social jet lag was an independent and important factor for high cardiovascular risk. Indeed, for every additional hour of social jet lag experienced, cardiovascular risk increased by >30%![14]

It is not only adults, however, who should worry about the effects of social jet lag on cardiovascular and cerebrovascular health. In a recent study, over 300 children aged between 8 and 10 years were examined for risk factors for cardiometabolic disease such as carbohydrate metabolism, blood pressure, cholesterol levels, and vascular health and associated with the extent to which each participant experienced social jet lag.[15] Neither sleep disturbances nor sleep duration were significantly associated with carbohydrate metabolism, blood pressure, cholesterol levels, and vascular health. However, social jet lag was significantly linked to both elevated cholesterol levels and reduced vascular health. These results suggest that the effects of social jet lag may be a substantial and assessable public health target for offsetting the negative trajectories for cardiometabolic disease risk in children.[15]

Another study of adults revealed that social jet lag increased several risk factors associated with cardiovascular disease. For example, experiencing moderate social jet lag decreased high-density lipoprotein-cholesterol levels, evoked higher triglyceride concentrations, and decreased insulin sensitivity.[16] Chronotype (a measure of preferred timing of sleep and wake-related activity) also seems to be linked to mortality from cardiovascular disease. Individuals with evening chronotypes (colloquially known as "night owls") are at a higher risk for mortality for all causes of death, but especially cardiovascular causes, than individuals with morning chronotypes (colloquially known as "morning larks").[17]

Even the relatively short time shift that occurs during the transition to Daylight Savings Time can affect cardiovascular function. The incidence of myocardial infarction and ischemic stroke hospitalizations is elevated the weekend following the springtime one-hour phase advance.[18] Considered together, social jet lag uncouples central and peripheral clocks, and impairs cardiovascular functioning, potentially increasing the risk of developing cardiovascular diseases.[19-20]

Exposure to Dim Lighting During the Day

Exposure to dim light during the day has the potential to alter risk factors associated with cardiovascular and cerebrovascular diseases. For example, exposure to dim light during the day has been associated with altered blood pressure regulation. Studies have reported that prolonged exposure to daytime dim light can lead to increased blood pressure, particularly during nighttime hours.[21] This disruption in blood pressure regulation may increase the risk of negative cardiovascular and cerebrovascular events and, if long-term, also contribute to the development of chronic hypertension.

Dim light exposure during the day has been reported to affect endothelial function. Endothelial function is a key factor in cardiovascular health. The endothelium is a thin membrane that lines the inside of the heart and other blood vessels. Endothelial cells release substances that control vascular relaxation and contraction, as well as enzymes that control blood clotting, immune function, and platelet (a colorless substance in the blood) adhesion. Endothelial dysfunction is characterized by impaired blood vessel dilation and increased inflammation, both of which are associated with an increased risk of cardiovascular and cerebrovascular diseases. Research suggests that exposure to dim light during the day may negatively affect endothelial function, with the potential to contribute to significant cardiovascular dysfunction.[22]

As noted in Chapter 4, exposure to dim light during the day can also affect sleep patterns, which in turn can influence cardiovascular function. Several studies have reported that exposure to dim light during the day may disrupt the natural sleep-wake cycle and impair the quality and duration of sleep.[23] Sleep disturbances, such as insufficient sleep or poor sleep quality, have been associated with an increased risk of cardiovascular diseases, including hypertension and heart disease.

To optimize cardiovascular health and mitigate the potential negative effects of dim light exposure during the day, consider the following points that we keep returning to:

1. Seek bright light exposure: Increase exposure to bright natural light during the daytime to support the regulation of the circadian rhythm and promote cardiovascular health.[24]
2. Maintain a regular sleep schedule: Establish a consistent sleep routine with adequate exposure to natural light during the day and dim light in the evening to support proper sleep-wake cycles and improve cardiovascular function.[22]
3. Limit exposure to dim light: Minimize exposure to prolonged periods of dim light during the day, especially in environments where bright light is available.

Exposure to Light at Night

Initially, many cardiovascular and cerebrovascular patients are admitted to an intensive care unit (ICU) prior to moving to a typical hospital room. It is probably not surprising that atypical sleep is often reported by patients and observed by medical staff in ICUs given the numerous disruptive factors commonly experienced at night, including light, noise, pain, administration of medication, and routine patient care activities such as phlebotomy and vital sign monitoring.[25] In addition, light intensities of up to 1,000 lux at night (the equivalent of an overcast day or typical TV studio lighting) have been reported in ICU settings![26] The lighting environment is obviously problematic for a healing environment, as one recent study reported that exposure to as little as 5–10 lux of light for just two nights significantly affected sleep architecture.[27] Several polysomnography studies have confirmed these observations. One recent study revealed that ICU patients and nonpatients spend comparable amounts of time asleep per day, however, the timing of sleep is altered in ICU patients. Among ICU patients, sleep is typically bifurcated into approximately equal amounts during the day and night, as opposed to occurring primarily at night.[28] Not surprisingly, ICU patients tend to display prolonged sleep latency (i.e., difficulty falling asleep). These patients also display substantial sleep fragmentation (multiple arousals), decreased sleep efficiency (ratio of time spent asleep to time in bed), elevated stage 2 sleep ("light sleep"), decreased stage 3 ("slow-wave" or "deep") sleep, and decreased rapid eye movement (REM) sleep.[29–31] Considered together, ICU patients display atypical sleep architecture and experience poor sleep quality, which in turn can impair recovery and precipitate other physiological, cognitive, and behavioral disorders commonly associated with disrupted or insufficient sleep including cardiovascular disorders.[32]

Indeed, the results of a clinical trial involving cardiac patients suggest that providing enhanced daytime brightness and restricting nocturnal light exposure improves sleep over the course of the hospital stay as compared to standard hospital room lighting.[33]

Less has been reported in the scientific literature about potential cardiovascular effects of exposure to light at night among individuals who are not night-shift workers. However, one home-based study with 528 elderly Japanese people demonstrated a significant increase in night-time systolic and diastolic blood pressure among individuals whose bedroom had nighttime light levels >5 lux.[34] Another home study of healthy young adults compared the effects of dark night-time environments to exposure to light at night (1,000 lux) on cardiovascular and respiratory function. Both the apnea-hypopnea index and the ratio of low-frequency power to high-frequency power (suggesting altered cardiovascular sympathetic control) were elevated in the study participants in light at night conditions compared to dark nights.[35] Similarly, a small sleep lab study that compared blood pressure among individuals exposed to dim light (<10 lux) versus bright light (>3,000 lux) at night reported that an elevation of systolic blood pressure among the bright light group emerged during the night.[36] The elevated blood pressure was reduced by treatment with melatonin supplements.

The use of 24-hour naturalistic lighting in rehabilitation units post-stroke indicates that patients experienced improved sleep with decreased disturbances, improved cognitive function, and stabilized endocrine function.[37] Another study of stroke recovery used artificial sunlight exposure therapy during the day, with increasing morning blue-spectral illuminance peaking in the afternoon to imitate daylight. Within four weeks of the occurrence of a stroke, and after a minimum of 14 days of this treatment that more closely mimicked the solar day, patients reported improved daily function.[38] Thus, normalization through resetting circadian rhythms to standard light-dark cycles post-stroke with the use of controlled lighting in a hospital setting is a noninvasive intervention that can potentially improve recovery rates.

Together, these studies indicate that many of the detrimental cardiovascular effects that have been so well-characterized in night-shift workers (see below), can emerge among non-night-shift workers who are merely exposed to light at night, sometimes even in the absence of altered sleep. The rapid onset of the cardiovascular and cerebrovascular effects in response to disrupted circadian rhythms may have particularly serious implications for patients experiencing nighttime light exposure in an ICU or at-risk individuals in their homes.

Experimental studies with animals can help provide direct evidence of the effects of disrupted circadian rhythms on outcomes from cardiovascular failures such as cardiac arrest or ischemic stroke. As noted, hospital patients may be particularly vulnerable to the consequences of light at night because of their compromised physiological state. In one mouse study, cardiac arrest with cardiopulmonary resuscitation was used to test the hypothesis that exposure to dim light at night impairs central nervous system recovery from a major pathological insult. To mimic the clinical conditions that humans experience after suffering a cardiac arrest and being assigned to the ICU for a week of around-the-clock light exposure, mice were exposed to dim light at night (5 lux) for five nights after experiencing cardiac arrest and resuscitation. One of the most striking results from this study was that after only one week of light at night exposure, usually considered to be a relatively innocuous environmental factor, mortality post–cardiac arrest nearly doubled compared to animals housed in dark nights (0 lux).[39] Neuronal damage in the brain, especially in the hippocampus, was significantly higher in surviving mice exposed to dim light at night for seven nights after cardiac arrest compared to those housed in dark nights. It was thought that exposure to light at night may have increased neuronal damage by promoting pro-inflammatory pathways in the brain; pro-inflammatory cytokine gene expression was elevated in mice exposed to light at night after the cardiac arrest. Cytokines are a broad and loose category of proteins important in cell signaling within the immune system. Because of their relatively large size, cytokines cannot cross the lipid membrane bilayer of cells to enter the cytoplasm and therefore typically exert their effects by interacting with specific cytokine receptors on the target cell surface. In many ways, cytokines are the hormones of the immune system. Immune responses to damaged brain tissues are not well regulated, and inflammation can impair neuronal cell function and even cause cell death. It appears that the elevation in cytokines in response to light at night directly provokes the negative consequences. Selective inhibition of the inflammatory cytokines, IL-1β or TNFα, ameliorated brain damage and survival rates in animals that experienced a cardiac arrest, then were exposed to dim light at night for seven days. The effects of light at night on the outcomes of cardiac arrest, including mortality, were also prevented by using long wavelength (red) light at night instead of a full spectrum (white) light source. As you now know, long wavelength (reddish) light has minimal effects on the endogenous circadian clock.

Similar studies using mouse models of ischemic stroke yielded similar results. A mouse model of stroke (middle cerebral artery occlusion) was used to test the hypothesis that exposure to dim light at night impairs recovery from a major stroke. Exposure to light at night for seven days after experiencing a stroke resulted in larger areas of brain damage compared to the size of post-stroke brain damage in animals housed in dark night conditions.[40] Additionally, animals experiencing strokes and then exposed to light at night during the subsequent week also displayed reduced survival, increased post-stroke anxiety-like behaviors, and elevated gene expression of pro-inflammatory cytokines in the injured hemisphere. Both exposure to dim red light at night or to environments in which the night-time lighting had the short wavelength (blue) light filtered out ameliorated the effects on light at night exposure on neuroinflammation, brain damage, behavioral changes, and survival.[40] Taken together, these data from animal studies suggest a low-cost environmental adjustment in lighting in the ICUs that might enhance outcomes: namely, filter out the blue light!

Night-Shift Work

In 1984, one of the first reviews suggesting an association between night-shift work and heart disease was published.[41] In the intervening 40 years, there have been many studies reporting an association between night-shift work and cardiovascular disease, virtually all of them coming to the same conclusion; namely, there is a strong association between night-shift work and cardiovascular disease.[42,43] There is also evidence that individuals who have worked night shifts for six or more years are at increased risk of developing cardiovascular and cerebrovascular diseases.[44] Taken together, both clinical and preclinical (animal) studies provide strong evidence that disrupted circadian rhythms initiate, maintain, or worsen cardiovascular disorders. The interaction of sleep disorders with disrupted circadian rhythms is critical for cardiovascular and cerebrovascular problems, as suggested in Chapter 4.

Night-shift workers display activity patterns that are typically misaligned with their internal clock, and are often exposed to relatively bright light at night, causing additional disruption of the circadian rhythms. Importantly, night-shift workers have up to a 40% increased risk of cardiovascular diseases![45] In order to investigate the effects of night-shift work on biomarkers of cardiovascular diseases, one study exposed long-term night-shift workers to one of two 3-day protocols; one protocol simulated night-shift work, whereas the other protocol simulated day-shift work. Simulated night shifts were significantly correlated with elevated blood pressure and increased C-reactive protein, an inflammatory blood marker that has been implicated in the progression of heart disease.[46,47] Exposure to similar protocols has provoked significant increases in other inflammatory cytokine markers in the blood such as IL6 and TNFα that are also considered risk factors for cardiovascular diseases.[46]

As noted earlier in the chapter, blood pressure "dipping" ($\geq$10% reduction in blood pressure as compared to daytime blood pressure readings) during sleep is a typical observation among healthy individuals, whereas "nondippers" (<10% reduction in blood pressure as compared to daytime levels) have been reported to be at risk for increased mortality and cardiovascular diseases.[48,49] One study demonstrated a potential link between night-shift schedules and blood pressure "dipping."[50] The study tested newly hired bus drivers who began working the early-morning shift (4:00 a.m.) and who were considered "dippers" prior to employment. After 90 days of working the early morning shifts, 62% had converted to "nondipper" status. Individuals who were hired at the same time, but worked a typical day shift all displayed typical, healthy blood pressure dipping after 90 days of employment.[50]

A systematic review and meta-analysis examined the association between night-shift work and cardiovascular events. The analysis revealed that night-shift work was associated with a significantly increased risk of both heart attacks and ischemic strokes.[51] The analysis included data from various studies involving different populations and provided robust evidence of the link between night-shift work and cardiovascular diseases.

Another comprehensive review of epidemiological data investigated the relationship between night-shift work and chronic diseases, including cardiovascular diseases.[52] The review highlighted that night-shift work was consistently associated with an increased risk of cardiovascular diseases, including hypertension, coronary heart disease, and stroke. The analysis included various observational studies and provided substantial evidence supporting the detrimental impact of night-shift work on cardiovascular health.

Although men are at significantly greater risk for cardiovascular disease compared to women, night-shift work also influences risk of cardiovascular disease in women. One prospective study examined the relationship between night-shift work and the risk of coronary heart disease in women.[53] The results demonstrated that women who worked rotating night shifts had a significantly higher risk of developing coronary heart disease compared to women who did not work night shifts. This study followed a large cohort of female nurses over a 10-year period, providing valuable long-term evidence on the effects of night-shift work on cardiovascular health.

Similarly, another large-scale prospective study investigated the relationship between rotating night-shift work and the risk of coronary heart disease in women. The study included over 180,000 women followed for up to 24 years. The results revealed that women who worked rotating night shifts had a significantly higher risk of developing coronary heart disease compared to those who did not work night shifts. Finally, another study examined the association between rotating night-shift work and mortality among female nurses. The researchers analyzed data from over 74,000 nurses and reported that long-term night-shift work was associated with an increased risk of total and cause-specific mortality, including cardiovascular mortality.[54] Taken together, these prospective studies provide robust and compelling evidence on the long-term negative effects of night-shift work on cardiovascular health in women.

A study to investigate the association between night-shift work and the morbidity of cardiovascular diseases was conducted on men and women blue-collar workers. Night-shift workers displayed a higher prevalence of cardiovascular diseases compared to their day-shift counterparts.[55]

Since the first review to suggest an association between shift work and heart disease in 1984,[41] additional clinical and pre-clinical studies have reported associations between night-shift work and cardiovascular disease or cerebrovascular disease.[55,56] Remarkably, there is strong evidence that individuals who have worked night shifts for six or more years in their past remain at high risk of developing cardiovascular disease.[44] Disrupted circadian rhythms can have significant effects on cardiac function. Dysregulated circadian rhythms, such as those caused by night-shift work, social jet lag, light at night, or chronic sleep disorders, can have negative consequences on cardiovascular and cerebrovascular health.

References

1. Thosar SS, Butler MP, Shea SA. 2018. Role of the circadian system in cardiovascular disease. *Journal of Clinical Investigation*, 128: 2157–2167.
2. Meléndez-Fernández OH, Walton JC, DeVries AC, Nelson RJ. 2021. Clocks, rhythms, sex, and hearts: how disrupted circadian rhythms, time-of-day, and sex influence cardiovascular health. *Biomolecules*, 11: 883.
3. Massin MM, Maeyns K, Withofs N, Ravet F, Gérard P. 2000. Circadian rhythm of heart rate and heart rate variability. *Archives of Disease in Childhood*, 83: 179–182.
4. Millar-Craig MW, Bishop CN, Raftery E. 1978. Circadian variation of blood-pressure. *Lancet*, 1: 795–797.
5. Ahmad FB, Cisewski JA, Xu J, Anderson RN. 2023. 2022 provisional mortality data: United States. *MMWR Morbidity and Mortality Weekly Report*, 72: 488–492.
6. https://www.who.int/news-room/fact-sheets/detail/the-top-10-causes-of-death.
7. Sewerynek E. 2002. Melatonin and the cardiovascular system. *Neuroendocrinology Letters*, 23 Suppl 1: 79–83.
8. Lapostolle F, Borron SW, Surget V, Sordelet D, Lapandry C, Adnet F. 2003. Stroke associated with pulmonary embolism after air travel. *Neurology*, 60: 1983–1985.
9. Staiger H. 2019. Melatonin and cardiovascular function. *Frontiers in Endocrinology*, 10: 87.
10. Youngstedt SD, Kline CE, Elliott JA, Zielinski MR, Devlin TM, Moore TA. 2009. Circadian phase-shifting effects of nocturnal exercise in older compared with young adults. *American Journal of Physiology*, 297: R845–R855.

11. Caliandro R, Streng AA, van Kerkhof LWM, van der Horst GTJ, Chaves I. 2021. Social jetlag and related risks for human health: a timely review. *Nutrients*, 13: 4543.

12. Sűdy ÁR, Ella K, Bódizs R, Káldi K. 2019. Association of social jetlag with sleep quality and autonomic cardiac control during sleep in young healthy men. *Frontiers in Neuroscience*, 13: 950.

13. Rutters F, Lemmens SG, Adam TC, Bremmer MA, Elders PJ, Nijpels G, Dekker JM. 2014. Is social jetlag associated with an adverse endocrine, behavioral, and cardiovascular risk profile? *Journal of Biological Rhythms*, 29: 377–383.

14. Madeira SG, Reis C, Paiva T, Moreira CS, Nogueira P, Roenneberg T. 2021. Social jetlag, a novel predictor for high cardiovascular risk in blue-collar workers following permanent atypical work schedules. *Journal of Sleep Research*, 30: e13380.

15. Castro N, Diana J, Blackwell J, Faulkner J, Lark S, Skidmore P, Hamlin M, Signal L, Williams MA, Stoner L. 2021. Social jetlag and cardiometabolic risk in preadolescent children. *Frontiers in Cardiovascular Medicine*, 8: 705169.

16. Wong PM, Hasler BP, Kamarck TW, Muldoon MF, Manuck SB. 2015. Social jetlag, chronotype, and cardiometabolic risk. *Journal of Clinical Endocrinology and Metabolism*, 100: 4612–4620.

17. Knutson KL, von Schantz M. 2018. Associations between chronotype, morbidity and mortality in the UK Biobank cohort. *Chronobiology International*, 35: 1045–1053.

18. Manfredini R, Fabbian F, Cappadona R, Modesti PA. 2018. Daylight saving time, circadian rhythms, and cardiovascular health. *Internal and Emergency Medicine*, 13: 641–646.

19. Dutheil F, Baker JS, Mermillod M, De Cesare M, Vidal A, Moustafa F, Pereira B, Navel V. 2020. Shift work, and particularly permanent night shifts, promote dyslipidaemia: a systematic review and meta-analysis. *Atherosclerosis*, 313: 156–169.

20. Grimaldi D, Carter JR, Van Cauter E, Leproult R. 2016. Adverse impact of sleep restriction and circadian misalignment on autonomic function in healthy young adults. *Hypertension*, 68: 243–250.

21. Morris CJ, Purvis TE, Hu K, Scheer FA. 2016. Circadian misalignment increases cardiovascular disease risk factors in humans. *Proceedings of the National Academies of Science (USA)*, 113: E1402–1411.

22. Figueiro MG, Rea MS, Pastorelli K, et al. 2013. Modification of the sleep/wake cycle and sleep architecture in the blind using morning short wavelength light. *Journal of Sleep Research*, 22: 507–516.

23. Chellappa SL, Steiner R, Oelhafen P, Lang D, Götz T, Krebs J, Cajochen C. 2013. Acute exposure to evening blue-enriched light impacts on human sleep. *Journal of Sleep Research*, 22: 573–580.

24. Wright KP, Jr, McHill AW, Birks BR, Griffin BR, Rusterholz T, Chinoy ED. 2013. Entrainment of the human circadian clock to the natural light-dark cycle. *Current Biology*, 23(16), 1554–1558.

25. Tembo AC, Parker V, Higgins I. 2013. The experience of sleep deprivation in intensive care patients: findings from a larger hermeneutic phenomenological study. *Intensive Critical Care Nursing*, 29: 310–316.

26. Meyer TJ, Eveloff SE, Bauer MS, Schwartz WA, Hill NS, Millman RP. 1994. Adverse environmental conditions in the respiratory and medical ICU settings. *Chest*, 105: 1211–1216.

27. Cho CH, Lee HJ, Yoon HK, Kang SG, Bok KN, Jung KY, Kim L, Lee EI. 2016. Exposure to dim artificial light at night increases REM sleep and awakenings in humans. *Chronobiology International*, 33: 117–123.

28. Pisani M. 2015. Sleep in the intensive care unit: an oft-neglected key to health restoration. *Heart Lung Journal of Acute Critical Care*, 44: 87.

29. Friese RS, Diaz-Arrastia R, McBride D, Frankel H, Gentilello LM. 2007. Quantity and quality of sleep in the surgical intensive care unit: are our patients sleeping? *Journal of Trauma Injury, Infection and Critical Care*, 63, 1210–1214.

30. Elliott R, McKinley S, Cistulli P, Fien M. 2013. Characterisation of sleep in intensive care using 24 h polysomnography: an observational study. *Critical Care*, 17: R46.

31. Freedman NS, Kotzer N, Schwab RJ, 1999. Patient perception of sleep quality and etiology of sleep disruption in the intensive care unit. *American Journal of Respiratory and Critical Care Medicine*, 159: 1155–1162.

32. Orzeł-Gryglewska J. 2010. Consequences of sleep deprivation. *International Journal of Occupational Medicine and Environmental Health*, 23: 95–114.

33. Giménez MC, Geerdinck LM, Versteylen M, Leffers P, Meekes GJ, Herremans H, de Ruyter B, Bikker JW, Kuijpers PM, Schlangen LJ. 2017. Patient room lighting influences on sleep, appraisal and mood in hospitalized people. *Journal of Sleep Research*, 26: 236–246.

34. Obayashi K, Saeki K, Iwamoto J, Ikada Y, Kurumatani N. 2014. Association between light exposure at night and nighttime blood pressure in the elderly independent of nocturnal urinary melatonin excretion. *Chronobiology International*, 31: 779–786.

35. Yamauchi M, Jacono FJ, Fujita Y, Kumamoto M, Yoshikawa M, Campanaro CK, Loparo KA, Strohl KP, Kimura H. 2014. Effects of environment light during sleep on autonomic functions of heart rate and breathing. *Sleep and Breathing*, 18: 829–835.

36. Burgess HJ, Sletten T, Savic N, Gilbert SS, Dawson D. 2001. Effects of bright light and melatonin on sleep propensity temperature, and cardiac activity at night. *Journal of Applied Physiology*, 91: 1214–1222.

37. Wang SJ, Chen MY. 2020. The effects of sunlight exposure therapy on the improvement of depression and quality of life in post-stroke patients: a RCT study. *Heliyon*, 6: e04379.

38. Durgan DJ, Crossland RF, Bryan RM. 2017. The rat cerebral vasculature exhibits time-of-day-dependent oscillations in circadian clock genes and vascular function that are attenuated following obstructive sleep apnea. *Journal of Cerebral Blood Flow and Metabolism*, 37: 2806–2819.

39. Fonken LK, Bedrosian TA, Zhang N, Weil ZM, DeVries AC, Nelson RJ. 2019. Dim light at night impairs recovery from global cerebral ischemia. *Experimental Neurology*, 317: 100–109.

40. Weil ZM, Fonken LK, Walker WH 2nd, Bumgarner JR, Liu JA, Melendez-Fernandez OH, Zhang N, DeVries AC, Nelson RJ. 2022. Dim light at night exacerbates stroke outcome. *European Journal of Neuroscience*, 52: 4139–4146.

41. Akerstedt T, Knutsson A, Alfredsson L, Theorell T. 1984. Shiftwork and cardiovascular disease. *Scandinavian Journal of Work and Environmental Health*, 10: 409–414.

42. Boggild H, Knutsson A. 1999. Shiftwork, risk factors and cardiovascular disease. *Scandinavian Journal of Work and Environmental Health*, 25: 85–99.

43. Mosendane T, Mosendane T, Raal FJ. 2008. Shift work and its effects on the cardiovascular system. *Cardiovascular Journal of Africa*, 19: 210–215.

44. Knutsson A, Akerstedt T, Jonsson B, Orth-gomer K. 1986. Increased risk of ischaemic heart disease in shift workers. *Lancet*, 8498: 89–92.

45. Hebl JT, Velasco J, McHill A. 2022. Work around the clock: how work hours induce social jetlag and sleep deficiency. *Clinical Chest Medicine*, 43: 249–259.

46. Morris CJ, Purvis TE, Mistretta J, Hu K, Scheer F. 2017. Circadian misalignment increases C-reactive protein and blood pressure in chronic shift workers. *Journal of Biological Rhythms*, 32: 154–164.

47. Castro AR, Silva SO, Soares SC. 2018. The use of high sensitivity C-reactive protein in cardiovascular disease detection. *Journal of Pharmacology and Pharmaceutical Sciences*, 21: 496–503.

48. Fagard RH, Celis H, Thijs L, Staessen JA, Clement DL, De Buyzere ML, De Bacquer DA. 2008. Daytime and nighttime blood pressure as predictors of death and cause-specific cardiovascular events in hypertension. *Hypertension*, 51: 55–61.

49. Ohkubo T, Hozawa A, Yamaguchi J, Kikuya M, Ohmori K, Michimata M, Matsubara M, Hashimoto J, Hoshi H, Araki T, Tsuji I, Satoh H, Hisamichi S, Imai Y. 2002. Prognostic significance of the nocturnal decline in blood pressure in individuals with and without high 24-h blood pressure: the Ohasama study. *Journal of Hypertension*, 20: 2183–2189.

50. McHill AW, Velasco J, Bodner T, Shea SA, Olson R. 2022. Rapid changes in overnight blood pressure after transitioning to early-morning shiftwork. *Sleep*, 45: zsab203.

51. Vyas MV, Garg AX, Iansavichus AV, Costella J, Donner A, Laugsand LE, Janszky I, Mrkobrada M, Parraga G, Hackam DG. 2012. Shift work and vascular events: systematic review and meta-analysis. *British Medical Journal*, 345: e4800.

52. Wang XS, Armstrong ME, Cairns BJ, Key TJ, Travis RC. 2011. Shift work and chronic disease: the epidemiological evidence. *Occupational Medicine*, 61: 78–89.

53. Kawachi I, Colditz GA, Stampfer MJ, Willett WC, Manson JE, Speizer FE, Hennekens CH.1995. Prospective study of shift work and risk of coronary heart disease in women. *Circulation*, 92: 3178–3182.

54. Gu F, Han J, Laden F, Pan A, Caporaso NE, Stampfer MJ, Kawachi I, Rexrode KM, Willett WC, Hankinson SE, Speizer FE, Schernhammer ES. 2015. Total and cause-specific mortality of U.S. nurses working rotating night shifts. *American Journal of Preventative Medicine*, 48: 241–252.

55. Abu Farha R, Alefishat E. 2018. Shift work and the risk of cardiovascular diseases and metabolic syndrome among Jordanian employees. *Oman Medical Journal*, 33: 235–242.

56. Boggild H, Knutsson A. 1999. Shiftwork, risk factors and cardiovascular disease. *Scandinavian Journal of Work and Environmental Health*, 25: 85–99.

8

STRATEGIES TO REDUCE DISRUPTED CIRCADIAN RHYTHMS AND IMPROVE HEALTH

This book has described the many different ways inappropriate exposure to light can compromise our health, such as affecting body weight, increasing risk for cancer and cardiovascular disease, and precipitating depression. I have reviewed how jet lag, social jet lag, night-shift work, exposure to dim light during the daytime, and exposure to light at night can disrupt our internal circadian rhythms. As suggested throughout the book, the best way to manage light exposure is to emulate what our Paleolithic ancestors experienced: maximize exposure to blue light early during the day and curtail exposure to blue light during the night. This chapter provides a summary of all the mitigation techniques for disrupted circadian rhythms described throughout the book. Some of the advice is overlapping for the various types of disruptors of circadian rhythms.

Dark Matters. Randy J. Nelson, Oxford University Press. © Oxford University Press (2025).
DOI: 10.1093/9780197639979.003.0008

In general, we can overcome the negative effects of disrupted circadian rhythms by increasing our exposure to short wavelength (blue) light during the early day and reducing our exposure to artificial light at night. In this final chapter I will suggest a number of the very low-tech (and a few high-tech) strategies for controlling exposure to light at night. Other popular timing (or chrono)-therapeutic strategies for preventing or reversing circadian rhythm disturbances include improving sleep hygiene, exogenous melatonin treatment, bright-light therapy, and shifting the timing of exercise and meals. Please note that improving sleep hygiene and improving circadian hygiene typically involve similar approaches. For example, a recent study of Brazilian adults concluded that to promote sleep quality and improve corresponding circadian health benefits, people should engage in regular exercise, preferably in the morning, as well as avoid naps, heavy meals close to bedtime, caffeine, smoking, and evening screen exposure.[1] Improving sleep timing by setting a consistent bedtime is recommended by the American Academy of Sleep Medicine, although there are few human studies on its effectiveness. Melatonin has been extensively used to treat jet lag, sleep disorders, and other circadian rhythm disturbances. Although melatonin improves the timing of when we fall asleep, there is little evidence that it influences how long and how well we sleep. It can also help to use blackout shades and sleep masks, and to cut back on the use of electronic devices, at least one, but ideally several, hours before bed.

Controlling the characteristics of light also makes a big difference. For example, using applications for computers, phones, and tablets that change the wavelength of the light they emit is helpful. Different kinds of light bulbs produce different spectrums, so one should choose an appropriate spectral distribution for the time of day they are using the light. Use broad-spectrum bulbs during the day and red-shifted lights at night. Although blue LED lights have been singled out as a big part of the problem with light at night, they can also be part of the solution if used during the day. One would then want to change to long-wavelength LED lights at night. An even more advanced technology allows you to program "smart" light bulbs to move from short to long wavelength light over the course of the day.

Throughout history, people were exposed to light levels manyfold higher than modern humans typically see indoors. Using brighter lighting indoors, or simply going outdoors in the morning, works well to synchronize your circadian rhythms. The following suggestions should help you attain optimal circadian hygiene under potentially challenging circumstances.

Jet Lag

As you know, jet lag occurs when our circadian rhythms are disrupted by rapid travel across different time zones. The following strategies can help alleviate the symptoms of jet lag and facilitate faster adjustment to the new time zone:

1. Gradually adjust your sleep-wake schedule: If possible, gradually shift your sleep and wake times prior to travel closer to the schedule of your destination. This can help prepare your body for the new time zone and minimize the disruption to your circadian rhythms.[2]

2. Stay hydrated: Drink plenty of water before, during, and after the flight. Proper hydration can help alleviate symptoms of fatigue and headaches associated with jet lag.[3]

3. Optimize exposure to natural light: Upon arriving, expose yourself to natural light during the daylight hours as much as possible. Natural light helps reset your internal clock and promotes adjustment to the new time zone.[4] Whenever I travel to Europe, the first thing on my agenda is to take a morning ride on the top section of an open-top tourist bus. These buses have no roof so that the people sitting on the top level can see better or in my case, be exposed to sunlight for an hour or so. Walking around achieves the same goal, and is what I do on subsequent days, but usually I'm too tired on the first day to walk for hours.

4. Minimize exposure to light at night: In the evening, reduce your exposure to bright lights, especially short wavelength (blue) light emitted by electronic devices at least three hours prior to sleep. As we've seen, blue light at night can suppress the production of melatonin, the hormone that regulates sleep-wake cycles. Minimizing light exposure at night can help promote better sleep and aid in the adjustment to the new time zone.[5,6]

5. Use melatonin supplements: If you live in the United States, then you can purchase melatonin supplements over the counter (OTC). However, in many countries melatonin is only available by prescription; this is a very important consideration when traveling internationally. Melatonin can help regulate your circadian rhythms and downstream rhythms such as your sleep-wake cycles to assist in adjusting to a new time zone. Consult with a healthcare professional

to determine the appropriate timing and dosage, and to ascertain that there are no contraindications for you,[7] and see important caveats below.

6. Stay active and exercise: Engage in regular physical activity during the day to help combat fatigue and promote better sleep at night. Exercise can also help regulate your body's internal clock and enhance adjustment to the new time zone.[8]

7. Consider strategic napping: Some consider strategic napping helpful in adapting to jet lag. Short naps of 20–30 minutes can help alleviate fatigue and enhance alertness during the day at your location. However, avoid napping close to bedtime to prevent disruption to nighttime sleep.[8]

Many people who travel multiple time zones often report the strategic use of melatonin to combat jet lag. Recall from Chapter 3 that research studies suggest the existence of an approximately 12-hour period during which treatment with a melatonin supplement advances circadian rhythms, and a 12-hour period during which melatonin treatment delays circadian rhythms. Thus, this information allows for the therapeutic use of melatonin in resetting the phase of circadian rhythms. For example, the advance zone is usually between 6 and 18 hours after awakening, whereas the delay zone usually begins about 18 hours after awakening and continues during sleep to about 6 hours after awakening. Thus, light and melatonin can be given to treat ailments such as jet lag, problems associated with shift work, advanced and delayed sleep phase disorders, and free-running rhythms among blind people. Recently, however, some in the sleep medicine field have been largely discouraging the use of OTC melatonin. Some of the reluctance is because in the United States, OTC melatonin is categorized as a dietary supplement and thus not subject to the same standards as medicine by the FDA. Years ago, melatonin supplements consisted of capsules of dried pineal glands that were obtained from cattle during slaughter. As assays for melatonin were developed, these capsules were tested, and it was discovered that these "melatonin" capsules contained little or no melatonin. As you might have already guessed, because most slaughterhouses operate primarily during the day, when the pineal gland produces little or no melatonin, the pineal glands harvested at this time were virtually devoid of this hormone. Supplement producers rushed to manufacture melatonin in their labs, and claimed much greater control over the amounts in these new supplements, typically 1, 3, or 5 mg. However, studies have reported that these stated doses are rarely accurate.[9] Presumably, the doses in the tablets/capsules are becoming harmonized with the labels in the intervening years.

People also often engage in the inappropriate use of melatonin. Because melatonin is often marketed as a natural, and therefore safe, product, the risk of overuse seems likely. If it works, then some folks tend to take more melatonin, a tactic that causes it not to work. In common with many substances, there is a tendency to habituate with overuse, requiring more of the substance. But with melatonin, higher dosing often leads to overwhelming the receptors and down-regulating them so they no longer respond.

Finally, the safety and efficacy of melatonin for children has not been determined. Annual pediatric melatonin overdoses rose from 8,000 in 2012 to more than 52,000 in 2021; 15% of children with overdoses were hospitalized.[10] Until access and better prevention of melatonin overdose in children are assured, it is best to avoid pediatric melatonin supplements.

Again, it is important to note that individual responses to jet lag may vary, and these strategies may not completely eliminate all symptoms. For personalized advice and specific considerations, I always recommend consultation with a healthcare professional.

Social Jet Lag

Social jet lag refers to the misalignment between individuals' biological clock and their social or work schedule, typically experienced during weekends or days off when sleep and wake times differ significantly from those on workdays. This temporal disruption can lead to symptoms similar to those experienced with travel-related jet lag, and is much more common than jet lag. Social jet lag can contribute to several health problems. As discussed throughout the book the following strategies can help mitigate the effects of social jet lag:

1. Maintain a consistent sleep schedule: Establish a regular sleep routine by going to bed and waking up at the same time every day, including weekends. Consistency in sleep patterns helps regulate your internal circadian clock and reduces the effects of social jet lag on health. Indeed, the effects of social jet lag phase shifts greater than four hours are much worse than phase shifts less than two hours.[11]

2. Create a sleep-friendly environment: Ensure that your bedroom is conducive to quality sleep. Keep the room dark,

quiet, and at a comfortable temperature. Use blackout curtains or an eye mask to block out external light, and consider using earplugs or white noise machines to minimize noise disruptions.[12]

3. Limit caffeine and alcohol intake: Avoid consuming caffeine close to bedtime, as it can interfere with sleep. Similarly, although alcohol may initially induce drowsiness, it can disrupt sleep patterns and lead to poor-quality sleep. Limit or avoid both substances, especially in the hours leading up to bedtime.[13]

4. Practice good sleep hygiene: Establish a presleep routine to signal your body that it is time to wind down. This may include activities such as reading, taking a warm bath, or practicing relaxation techniques like deep breathing or meditation. Engaging in relaxing activities before bed can facilitate the transition into restful sleep.[14]

5. Avoid electronic devices before bed: The blue light emitted by electronic devices such as smartphones, tablets, and computers can suppress the production of melatonin and interfere with sleep. If you have iPads, iPhones, or other Apple products, then you can simply activate "Night Shift" on your device. Go to "Settings," then navigate to "Display and Brightness." Now simply toggle Night Shift mode on. You can also set specific times for this to occur automatically each day. For other types of devices, you can get one of several apps that are available. Among the best blue light reducing apps are F.LUX and Twilight. Even with these apps, you should try to minimize or avoid using electronic devices at least one hour before bedtime as they do not perfectly screen out light that might disrupt your

circadian rhythms.[15] Recent studies[16] reported minimal or no effects of the "Night Shift" app on sleep, but the control group (no phone use at night) did not display differences in sleep quality or time to fall asleep, in contrast to other studies showing the effects of light exposure on the onset of melatonin secretion and sleep.[17]

6. Consider a short nap: If you experience fatigue or drowsiness due to social jet lag, then a short nap of 20–30 minutes during the day can provide a temporary energy boost. However, avoid napping too close to your regular bedtime, as it may interfere with nighttime sleep.[18]

7. Seek daylight exposure: Expose yourself to natural daylight during the day, especially in the morning. Natural light helps regulate your circadian rhythms and promotes alertness, making it easier to adjust to your desired sleep schedule.[16]

Again, individual responses to social jet lag may vary. If you experience significant difficulties with sleep, using sleep masks, or excessive daytime sleepiness, then it is advisable to consult with a healthcare professional for further evaluation and personalized recommendations.

Exposure to Dim Lighting During the Day

The advice for ameliorating the effects of this disruptor of circadian rhythms is relatively straightforward. Exposure to bright light during the daytime is crucial for maintaining healthy circadian rhythms, which are critical for regulation of our sleep-wake and other physiological and behavioral rhythms. When daylight is insufficient to synchronize our rhythms, then they may lose temporal harmony and our health can suffer. The following strategies can help mitigate the effects of dim light on circadian rhythms:

1. Increase exposure to natural daylight: Spend time outdoors during daylight hours, especially in the morning. Exposure to natural light, particularly in the early part of the day, helps regulate the body's internal clock and promotes alertness.[19]

2. Use bright artificial light: If natural daylight is limited, then consider using bright artificial light sources indoors. Light therapy devices used in the treatment of SAD, or specialized light bulbs that emit especially bright light (at least 2,000 lux for most people), preferably with a color temperature close to natural daylight, can help simulate the effects of sunlight.[20] Use artificial lights that provide more than 20 $\mu W/cm^2$ of sky-blue light during the day.[21]

3. Maintain a well-lit environment: Ensure that your workspace and living areas are adequately lit during the daytime with as much natural light as possible. Open curtains or blinds to allow natural light to enter your space, and use bright indoor lighting to compensate for dim light conditions.[22]

4. Limit exposure to dim light sources: Minimize exposure to dim light sources, such as dimly lit rooms, low-intensity lamps, or electronic devices with low brightness settings, especially during the daytime. Dim light can interfere with the body's internal clock and suppress the release of alertness-promoting hormones.[19]

5. Take regular breaks: If your work or living environment involves prolonged exposure to dim light, take regular breaks to step into well-lit areas or go outdoors. This can help counteract the effects of dim light on your circadian rhythms and improve alertness.[19]

6. Optimize the sleep environment: Create a sleep environment that is conducive to darkness at night. Use blackout curtains, wear an eye mask, or eliminate sources of artificial light in the bedroom to promote better sleep quality and support the natural synchronization of your circadian rhythms.[21]

If you experience persistent disruptions to your circadian rhythms due to dim light exposure or have underlying sleep-related or mental health issues, then it is advisable to consult with a healthcare professional who can provide personalized advice and recommendations.

Exposure to Light at Night

Exposure to light at night, particularly artificial light, can disrupt our circadian rhythms and have long-term negative effects on our health. Here are some strategies to mitigate the impact of light at night:

1. Create a sleep-friendly environment: Make your bedroom conducive to sleep by ensuring it is dark, quiet, and cool. Use blackout curtains, blinds, or a sleep mask to block out external light sources. On some electronic devices, such as alarm clocks, it is possible to shift the color of the display

to red from blue. If not possible, then cover or dim any electronic devices that emit light, such as alarm clocks or standby lights.[22]

2. Limit electronic device use for at least three hours before bed: The blue-enriched light emitted by electronic devices, such as smartphones, tablets, and computers, can suppress the production of melatonin, a hormone that regulates circadian rhythms and the onset of sleep. Limit blue wavelengths to less than $2\mu W/cm^2$ during the evening. Avoid using these devices at least 2–3 hours before bedtime to allow your body to prepare for sleep.[21-25]

3. Use dim and warm lighting in the evening: Switch to low-intensity, warm-colored lighting in the evening to create a relaxing atmosphere. Avoid bright and cool-colored lights, as they can mimic daylight and interfere with circadian rhythms including your natural sleep-wake cycles.[26]

4. Install blue light filters: Consider using blue light filters or apps that reduce the amount of blue light emitted by electronic devices. These filters can help minimize the impact of blue light on melatonin production and improve sleep quality.[26] There are also glasses with blue-light filtering lenses that can be purchased online.

5. Minimize exposure to bright light at night: Limit exposure to bright lights, especially during the late evening and throughout the night. Bright light can disrupt circadian rhythms, suppress melatonin production, and disrupt sleep. Use low-intensity lighting or dimmers in your home during the nighttime hours.[27,28]

6. Establish a consistent sleep schedule: Maintain a regular sleep routine by going to bed and waking up at the same

time every day, even on weekends. Consistency in sleep patterns helps regulate the body's internal clock and promotes better sleep quality.[29]

7. Consider using blackout curtains: If you live in an area with relatively bright external night-time light sources, consider using blackout curtains or shades in your bedroom. These can effectively block out external light, preserving circadian rhythmicity, and create a darker sleeping environment.[29]

It is critical to note that individual responses to light exposure vary, and these strategies may not be suitable for everyone. If you have persistent sleep problems or concerns, then you should consult with a healthcare professional for personalized advice and recommendations.

Night-Shift Work

Finally, shift work represents a unique and profound form of circadian disruption.[29] Shift workers have less flexibility in implementing the behavioral approaches listed above because of the need to adhere to unusual and fluctuating schedules. It has therefore been proposed that, for night-shift workers, taking factors such as chronotype (whether people identify as "night owls" or "morning larks") into consideration could prove to be critical in alleviating circadian disruption and improving sleep. Implementing some of these simple solutions may help to overcome the negative effects of circadian disruption on health and physiology.

As mentioned, what likely makes night-shift work so detri
mental to health is that night-shift individuals experience many
factors working against their circadian rhythms. Night-shift
workers experience increased exposure to light, elevated caloric
intake, and reduced sleep during their "night," which usually is
attempted in a relatively bright bedroom illuminated through
the blinds or curtains by the sun. Additional bad news for night-
shift workers is that they tend to eat high calorie food and display
low levels of activity relative to individuals who perform the
same jobs during the day shift.

Night-shift work, characterized by working during the night
and sleeping during the day, can dramatically disrupt circadian
rhythms and have various negative effects on physical and mental health. Even for night-shift workers, adopting a Paleo lighting
routine, but in reverse, will help. Put simply, make day your
night and make night your day in terms of lighting. Here are
some strategies to minimize the effects of night-shift work on
health:

1. Establish a consistent sleep schedule: Maintain a regular
   sleep schedule, even on days off. Try to sleep in a cool, dark,
   and quiet environment to promote quality sleep during
   the day.[30]
2. Optimize the sleep environment: Use blackout curtains or
   a sleep mask to create a dark sleeping environment that
   mimics nighttime conditions. Consider using earplugs or
   white noise machines to block out daytime noise and dis-
   tractions.[29]
3. Maintain a healthy diet: Focus on a well-balanced diet
   that includes nutritious foods. Avoid heavy meals before

bedtime, as they can interfere with sleep. Stay hydrated by drinking plenty of water throughout the shift.[30,31]

4. Limit caffeine and stimulant intake: Minimize or avoid consuming caffeinated beverages or stimulants close to bedtime, as they can disrupt sleep. Be cautious with the timing and quantity of these substances to prevent interference with your ability to fall asleep during the day.[32]

5. Promote physical activity: Engage in regular exercise or physical activity during your waking hours, especially right before your night shift begins. Exercise can improve sleep quality, increase alertness, and provide overall health benefits.[33]

6. Optimize light exposure: Prioritize exposure to bright light during your night shift to promote alertness and reduce drowsiness. There are glasses on the market to treat SAD that can provide you with bright light while not affecting others (e.g., night-shift nurses). Use bright lighting in your workspace.[34,35]

7. Seek social support: Maintain open communication with your family, friends, and colleagues about the challenges of night-shift work. Seek support from others who understand your unique work schedule and its impact on your health and well-being.[31]

8. Consider rotating shifts clockwise: If you must work a rotating shift, then request a clockwise rotation of shifts (morning to evening to night) rather than counterclockwise. This allows for easier adjustment to a new schedule and aligns better with your free-running circadian rhythms.[34]

9. Use blue-light-blocking glasses during the day: You want to avoid blue light during the day, as it will reset your circadian rhythms. If possible, during your commute home and anytime that you are outdoors or exposed to bright indoor lighting during the day (your night), avoid short wavelength (blue) light.[36]

10. Pay attention to the timing of food intake: Fast between dinner and breakfast, preferably 14 or more hours. This is especially true for night-shift workers. It is hypothesized that fasting and time-restricted feeding regimens that actively impose a diurnal rhythm of food intake that is aligned with the 24-hour solar day improves daily cycles in circadian clock gene expression, reprogramming of molecular mechanisms of energy metabolism, improved insulin metabolism, and improved body weight regulation.[37]

If you are a night-shift worker, it is prudent to consult with your employer, occupational health services, or a healthcare professional to discuss any specific concerns related to your individual health and work requirements.

Current and future research on the effects of disrupted circadian rhythms is focused on simple behavioral modifications to improve overall health. As we've learned from several strategies to improve health via behavioral changes, such as substance use disorders, diet and exercise, and safe sexual practices, behavioral modification is the most difficult strategy for improving outcomes. People generally seek a simple solution such as a pill to achieve their optimal health status. Nonetheless, for motivated individuals it is relatively easy to make a few lifestyle changes to improve the odds of good health.

* * * * *

The primary goal of this book was to bring readers a heightened awareness of the health and well-being consequences of poor circadian hygiene and to provide suggestions for strategies that may counter any negative health effects. The chapters in this book summarized the most reliable research on the different ways that exposure to light at night or lack of light during the day-time hours can affect our physiology, behavior, and ultimately, our health. As we've seen across the dozens of studies cited in this book, exposure to artificial ("blue") light at night can derail our circadian clock and wreak havoc on the temporal coordination of our physiology and behavior, leading to dysfunctional metabolism and cell division cycles and compromising immune, brain, and heart function. There is strong consensus among circadian biologists that robust circadian rhythmicity is critical to good health, that disrupting circadian rhythms can impair health, that regular daily exposure to sunlight can enhance night-time sleep, and that exposure to blue-enriched (480 nm) light at night can disrupt circadian rhythms and health.[38]

Jet lag, social jet lag, exposure to light at night and night shift work all can expose us to lighting schedules that lead to temporal disorganization. Similarly, missing out on bright full-spectrum light during the day will also derange our circadian rhythms, increasing our risk for several disorders that can negatively affect health. Nonetheless, the simple risk-reduction strategies that we have discussed and revisited in different contexts throughout the book can be used by anyone to address problematic light exposure. Remember that dark (and light) matters!

References

1. Castro-Santos L, Lima MO, Pedrosa AKP, Serenini R, de Menezes RCE, Longo-Silva G. 2023. Sleep and circadian hygiene practices association with sleep quality among Brazilian adults. *Sleep Medicine*, 6: 100088.
2. Waterhouse J, Reilly T, Atkinson G, et al. 2007. Jet lag: trends and coping strategies. *Lancet*, 369: 1117–1129.
3. Reilly T, Waterhouse J, Edwards B. 2005. Jet lag and air travel: implications for performance. *Clinical Sports Medicine*, 24: 367–380.
4. Eastman CI, Burgess HJ. 2009. How to travel the world without jet lag. *Sleep Medicine Clinics*, 4: 241–255.
5. Challet E. 2015. Keeping circadian time with hormones. *Diabetes, Obesity and Metabolism*, 17(Suppl 1): 76–83.
6. Herxheimer A, Petrie KJ. 2002. Melatonin for the prevention and treatment of jet lag. *Cochrane Database Systematic Reviews*, 2: CD001520.
7. Driver HS, Taylor SR. 2000. Exercise and sleep. *Sleep Medicine Reviews*, 4: 387–402.
8. Hayashi M, Masuda A, Horiuchi S, et al. 2016. The impact of napping on the association of sleep duration with mortality: a prospective cohort study. *Sleep*, 39: 315–322.
9. McFadden E, Jones ME, Schoemaker MJ, Ashworth A, Swerdlow AJ. 2014. The relationship between obesity and exposure to light at night: cross-sectional analyses of over 100,000 women in the Breakthrough Generations Study. *American Journal of Epidemiology*, 180: 245–250.
10. Grigg-Damberger MM, Ianakieva D. 2017. Poor quality control of over-the-counter melatonin: what they say is often not what you get. *Journal of Clinical Sleep Medicine*, 13: 163–165.
11. Wittmann M, Dinich J, Merrow M, et al. 2006. Social jetlag: misalignment of biological and social time. *Chronobiology International*, 23: 497–509.
12. Medic G, Wille M, Hemels ME. 2017. Short- and long-term health consequences of sleep disruption. *Nature and Science of Sleep*, 9: 151–161.

13. Roehrs T, Roth T. 2008. Caffeine: sleep and daytime sleepiness. *Sleep Medicine Reviews*, 12: 153–162.

14. Kline CE. 2014. The bidirectional relationship between exercise and sleep: implications for exercise adherence and sleep improvement. *American Journal of Lifestyle Medicine*, 8: 375–379.

15. Chang AM, Aeschbach D, Duffy JF, et al. 2015. Evening use of light-emitting eReaders negatively affects sleep, circadian timing, and next-morning alertness. *Proceedings of the National Academies of Science (USA)*, 112(4): 1232–1237.

16. Moore-Ede, M, Blask DE, Cain SW, Heitmann A, & Nelson RJ. 2023. Lights should support circadian rhythms: evidence-based scientific consensus. *Frontiers in Photonics*: 4. https://doi.org/10.3389/fphot.2023.1272934.

17. Duraccio KM, Zaugg KK, Blackburn RC, Jensen CD. 2021. Does iPhone night shift mitigate negative effects of smartphone use on sleep outcomes in emerging adults? *Sleep Health*, 7: 478–484.

18. Hayashi M, Masuda A, Horiuchi S, et al. 2016. The impact of napping on the association of sleep duration with mortality: a prospective cohort study. *Sleep*, 39: 315–322.

19. Cajochen C, Chellappa SL, Schmidt C. 2014. Circadian and sleep-wake dependent aspects of subjective alertness and cognitive performance. *Journal of Sleep Research*, 23: 364–374.

20. Revell VL, Burgess HJ. 2011. Bright light therapy: seasonal affective disorder and beyond. *Open Neuroendocrinology Journal*, 4: 109–120.

21. Moore-Ede, M. 2021. LEDs must spectrally balance illumination, circadian health, productivity, and energy efficiency. *LED Magazine*, www.eedsmagazine.com/lighting-health-wellbeing/article/14199941/ideal-led-lightingmust-balance-multiple-objectives-magazine.

22. Smith KA, Schoen MW, Czeisler CA. 2004. Adaptation of human pineal melatonin suppression by recent photic history. *Journal of Clinical Endocrinology and Metabolism*, 89: 3610–3614.

23. Rüger M, Scheer FAJL. 2009. Effects of circadian disruption on the cardiometabolic system. *Reviews in Endocrinology and Metabolic Disorders*, 10: 245–260.

24. Lee K, Barratt MJ, Yee BJ, et al. 2017. Effects of bright light treatment on subjective and objective measures of sleepiness during the daytime. *Sleep*, 40: zsw048.

25. Grandner MA, Hale L, Moore M, Patel NP. 2010. Mortality associated with short sleep duration: the evidence, the possible mechanisms, and the future. *Sleep Medicine Review*, 14: 191–203.

26. Vandewalle G, Schmidt C, Albouy G, et al. 2007. Brain responses to violet, blue, and green monochromatic light exposures in humans: prominent role of blue light and the brainstem. *PLoS One*, 2: e1247.

27. Burkhart K, Phelps JR. 2009. Amber lenses to block blue light and improve sleep: a randomized trial. *Chronobiology International*, 26: 1602–1612.

28. Crowley SJ, Eastman CI. 2015. Phase advancing human circadian rhythms with morning bright light, afternoon melatonin, and gradually shifted sleep: can we reduce morning bright-light duration? *Sleep Medicine*, 16: 288–297.

29. Rajaratnam SMW, Arendt J. 2001. Health in a 24-h society. *Lancet*, 358: 999–1005.

30. Kecklund G, Axelsson J. 2016. Health consequences of shift work and insufficient sleep. *British Medical Journal*, 355: i5210.

31. Niedhammer I, Chastang JF, Gendrey L, David S, Degioanni S. 2008. Importance of psychosocial work factors on general health outcomes in the national French SUMER survey. *Occupational Medicine (London)*, 58: 15–24.

32. Drake C, Roehrs T, Shambroom J, Roth T. 2013. Caffeine effects on sleep taken 0, 3, or 6 hours before going to bed. *Journal of Clinical Sleep Medicine*, 9: 1195–1200.

33. Samuels C. 2005. Sleep, recovery, and performance: the new frontier in high-performance athletics. *Neurological Clinics*, 23: 1107–1118.

34. Costa G. 2003. Shift work and occupational medicine: an overview. *Occupational Medicine (London)*, 53: 83–88.

35. Ruggiero JS, Redeker NS. 2014. Effects of napping on sleepiness and sleep-related performance deficits in night-shift workers: a systematic review. *Biological Research in Nursing*, 16: 134–142.

36. Aarts MPJ, Hartmeyer SL, Morsink K, Kort HSM, de Kort YAW. 2020. Can special light glasses reduce sleepiness and improve sleep of nightshift workers? A placebo-controlled explorative field study. *Clocks and Sleep*, 2: 225–245.

37. Lowe DA, Wu N, Rohdin-Bibby L, Moore AH, Kelly N, Liu YE, Philip E, Vittinghoff E, Heymsfield SB, Olgin JE, Shepherd JA, Weiss EJ. 2020. Effects of time-restricted eating on weight loss and other metabolic

parameters in women and men with overweight and obesity: the TREAT randomized clinical trial. *Journal of the American Medical Association, Internal Medicine*, 180: 1491–1499.

38. Rishi MA, Khosla S, Sullivan SS. 2023. Public Safety and the Public Awareness Advisory Committees of the American Academy of Sleep Medicine. Health advisory: melatonin use in children. *Journal of Clinical Sleep Medicine*, 19: 415.

INDEX

For the benefit of digital users, indexed terms that span two pages (e.g., 52–53) may, on occasion, appear on only one of those pages.

Tables, figures, and boxes are indicated by an italic *t, f, b* following the page number.

204